iPhone 12 U

The Complete New Guide to the iPhone 12 and iPhone 12 Pro Max, For Beginners and Seniors

Milton Don Randall

Copyright © 2022 by Milton Don Randall

This document is intended to provide accurate and dependable information about the subject and issues discussed. The publication is sold with the understanding that the publisher is not obligated to provide accounting, legally permissible, or otherwise qualified services. If legal or professional advice is required, a practicing member of the profession should be contacted.

From a Declaration of Principles that was unanimously accepted and approved by an American Bar Association Committee and a Publishers and Associations Committee. No portion of this document may be reproduced, duplicated, or transmitted electronically or in printed form. The recording of this book is expressly forbidden, and storage of this content is not permitted without the publisher's written consent. Each and every right is reserved.

The information contained herein is stated to be accurate and consistent, and any liability incurred as a result of inattention or otherwise as a result of the recipient reader's use or abuse of any policies, processes, or directions contained herein is sole and complete. Under no conditions will the publisher be held liable for any reparation, damages, or monetary loss incurred as a result of the information contained herein, either explicitly or implicitly.

All copyrights not held by the publisher are owned by the respective author(s).

The information contained herein is provided solely for informational purposes and is therefore universal. The information is presented without contract or assurance of any kind.

The trademarks are used without the trademark owner's consent, and the trademark is published without the trademark owner's permission or support. All trademarks and brands mentioned in this book are solely for clarity purposes and are owned by their respective owners, who are not affiliated with this document.

Free Bonus

Grab My *"Social Media Marketing Made Simple"* Ebook For **FREE!**

Today you can grab your copy of my Free e-book titled – **Social Media Marketing made Simple**. Best of all, it won't cost you a thing.

Click the image above to **Download the Book**, and also Subscribe for Free books, giveaways, and new releases by me.

https://mayobook.com/milton

My Other Book(s)

I recommend these books to you, it will be of help, check it out after reading this book.

1. **3D Printer:** A Complete 3D Printing Guide
2. **iPhone 13 User Guide**
3. **How to Win Playing Cribbage/Draw Poker Card Games:** The Ultimate Guide on Rules & Strategies to Win and Beat the Odds Playing Card Games Like a Pro
4. **Draw Poker Handheld Game:** The Ultimate Beginner's Guide on Rules & Strategies to Win and Beat the Odds Playing Poker Card Games Like a Pro
5. **iPad Pro**
6. **iPhone 8:** The Complete User Guide is for Beginners
7. **iPhone 11:** The Complete User Guide for Beginners
8. **iPhone 12 User Guide:** The Complete New Guide to the iPhone 12 and iPhone 12 Pro Max, For Beginners and Seniors

Table of Contents

IPHONE 12 USER GUIDE 1

FREE BONUS 4

MY OTHER BOOK(S) 5

INTRODUCTION 11

CHAPTER 1 14

   IPHONE SETUP 14
      iPhone Setup: The Basics 15
      Auto Setup for iPhone 16
      Restoring Data from back-up of old iPhone 17
      Setup Face ID 17
      Create iPhone Email 18
      Advanced iPhone Email Tweaks 20
      Protecting Your Online Activity 22
      iCloud is Everywhere: 23
      Services Subscription during iPhone Setup 24
      More Tweaks to the iPhone's Setup 25

CHAPTER 2 27

   IPHONE 12 OVERVIEW 27
   IPHONE 12 PRO OVERVIEW 28
   USING THE IPHONE 12'S SCREEN RECORDING APP 30
   HOW TO ADD SCREEN RECORD TO YOUR IPHONE 12 30
   SCREEN RECORDING ON IPHONE 12 31
      How to Screen Record with Sound on iPhone 12 33
      How to adjust the Settings for Screen Recording 34
      Limitations For Screen Recording 34

CHAPTER 3 36

   IPHONE 12 AND AIRPODS 36
   DOES THE IPHONE 12 COME WITH AIRPODS? 36
   THE IPHONE 12'S MISSING HEADPHONES 37
   THE IPHONE 12'S HEADPHONE OPTIONS 38

How to Shut down iPhone 12 39
    When the Basics Don't Work, Here's How to Shut Down Your iPhone 12 41
    How To Start The iPhone 12 41
    How to Restart iPhone 12 42
    How to Force-restart Your iPhone 12 43

CHAPTER 4 .... **45**
    IPHONE 12 AND IPHONE 12 PRO GUIDELINES 45

CHAPTER 5 .... **55**
    THE IPHONE 12 CAMERAS & VOICE MAIL 55
    AN OVERVIEW OF THE DATA 56
        Night Life 57
        ProRAW 58
    HOW TO CONFIGURE IPHONE 12 VOICEMAIL 59
    WHAT EFFECT DOES YOUR TELEPHONE SERVICE PROVIDER HAVE ON YOUR VOICEMAIL PREFERENCES? 62
    IS IPHONE VOICEMAIL LIKE VISUAL VOICEMAIL? 62
        How to Setup iPhone 12's Visual Voicemail Transcription 63
        Managing Your iPhone 12 Voicemail 64

CHAPTER 6 .... **66**
    HOW TO CUSTOMIZE IPHONE 66
        Personalize your iPhone text and ringtone sounds. 66
        Additional iPhone Customisation Options 67
        Customize iPhone Home screen 69
        iPhone Customizations that makes things Better to See 71
        iPhone Lock Screen Customization: 72
        Customize Notifications on Your iPhone 73

CHAPTER 7 .... **76**
    SIRI ON IPHONE 12 76
    HOW TO MAKE USE OF SIRI ON IPHONE 12 76
    HOW TO ENABLE SIRI ON IPHONE 12 76
    HOW TO MAKE USE OF SIRI ON IPHONE 12 77
    IOS 15 BRINGS SIRI SOME MUCH-NEEDED UPDATES 78
        How to Make use of Siri as an Intercom 80

CHAPTER 8 82

HOW TO TAKE SCREENSHOT ON IPHONE 12 82
HOW TO FIND YOUR IPHONE 12 SCREENSHOT 83
HOW TO SHARE IPHONE 12 SCREENSHOTS 84
HOW TO CLOSE APPS ON IPHONE 12 85
    When You Should Close iPhone Apps 87
    Close Non-Working Apps 88

CHAPTER 9 89

ITUNES, ICLOUD & ANDROID BACKUP RESTORE 89
HOW TO MOVE DATA FROM AN ANDROID PHONE 93
IPHONE BACKUP RESTORE FROM ICLOUD OR ITUNES 100

CHAPTER 10 106

HOW TO RESTART AN IPHONE 106
HOW TO RESTART THE NEW IPHONE 106
HOW TO RESTART DIFFERENT MODELS OF IPHONES 107
HOW TO PERFORM A HARD RESET ON THE IPHONE. 108
OTHER IPHONE MODELS: HOW TO PERFORM A HARD RESET. 109
HOW TO PERFORM A HARD RESET ON AN IPHONE 7 SERIES 110
MORE HELP RESETTING YOUR IPHONE 110

CHAPTER 11 112

HOW TO SHOW BATTERY PERCENTAGE ON AN IPHONE 12 112
ADDING A BATTERY WIDGET TO YOUR IPHONE 12 114
HOW TO MAKE USE OF APPLE PAY ON IPHONE 12 116
HOW TO MAKE USE OF APPLE PAY IN-STORE 118
    How to Modify Your Default credit Card on Apple Pay on iPhone 12 . 119

CHAPTER 12 121

IOS 15 121
WHAT IS THE IMPORTANCE OF IOS 15'S NEW PROTECTION FEATURES FOR YOUR IPHONE? 127
    Are there going to be fewer apps as a result of this? 129
    What does this all mean? 131
COMPLETE FIVE ACTIONS ON START OR SETUP WIRELESS CARPLAY ON IPHONE 132

CHAPTER 13 135

IPHONE 12 CHEAT SHEET 135
MORE FEATURES ON AN IPHONE 12 141
WHAT ARE THE DIFFERENCE BETWEEN THE FOUR IPHONE 12 VERSIONS? .. 142
  What are the Main Competitors of the iPhone 12? 144

CHAPTER 14 147

NEW IPHONE QUICK FIX 147
  iCloud Activation Lock Removal 148
  How to Remove Activation Lock on iPhone 149
  Using the Find My iPhone app, here's how to format an iPhone. 150
  How to Wipe an iPhone Using iCloud 151

CHAPTER 15 152

TOP RECOMMENDED IPHONE APPS 152

CHAPTER 16 166

HOW TO FORMAT IPHONE 12 166
  How to Restore and Reinstall, Clear iOS/Firmware on iPhone 12 168
  How to Unlock or Fix or Bypass Forgotten Security Password 168
  How to Make a Backup of Your iPhone 12 Data and Restore It 169
  How to Update iOS at iPhone 12 169
  How do you enter Accessibility mode on iPhone 12 and 12 Pro! 170

CHAPTER 17 171

IPHONE TIPS & TRICKS 171
  How to Enable USB Restricted Setting on iPhone 171
  Use Your iPhone to Control Your Apple TV. 172
  Make use of the Two Pane Scenery View option. 172
  Turning Off iPhone Alarms by with your face. 173
  Disable Face ID in a flash. 173
  How to Slow the two times click necessary for Apple Pay 174

CHAPTER 18 175

IPHONE 12 PROBLEMS & SOLUTION 175
HOW TO IMPROVE THE LIFE OF THE BATTERY ON AN IPHONE 12 176
HOW TO REPAIR IPHONE 12 MISSING 5G 183

How to Fix Wi-Fi Issues 184
    How to Fix Bluetooth Issues 186
    iPhone 12 Charging Problems: How to Fix It 187
    How to Fix Cellular Network Issues 188
    How to Fix Sound Issues 189
    How to Repair iPhone 12 Activation Problems 190
    How to Repair the Performance of an iPhone 12 191
    Stop Auto Downloads 195
    How to Repair Face ID Problems on an iPhone 12 198
    How to Fix Overheating Issues on an iPhone 12 199

FREE BONUS 201

FEEDBACK 202

ABOUT THE AUTHOR 204

MY OTHER BOOK(S) 206

INDEX 207

Introduction

Learn everything about the latest iPhones, including its features, design, specs, as well as user guide. A complete guide to everything there is to know about the brand new iPhone, the iPhone 12 and the iPhone 12 Pro Max, for both beginners and seniors.

The Apple iPhone is one of the most popular smartphones on the market today. If you're looking to get the best out of your device, there are many tips and tricks to help you get the most out of your new iPhone 12.

In this user guide, we explain all the things you need to know about the iPhone 12 and iPhone 12 Pro Max, including how they work, what's new and even tips on how to get the most out of your phone.

If you are planning to buy the iPhone 12 or iPhone 12 pro max then this guide will help you to understand everything about the iPhone 12 including all features, specifications, differences between iPhone 12 and iPhone 12 plus, and much more. This is a guide to help all beginners and seniors to understand the new iPhones and to use them well.

For a lot of people, it seems like the iPhone 11 is pretty good. The camera's got the same size sensor as the iPhone 11 Pro, and it still comes with a decent battery life. It's not perfect, but I think it's a good balance between performance and features. So if you're an iPhone 11 user, you're probably wondering why I'm recommending the iPhone 12. This guide is for you.

If you want the best iPhone 11 experience, the iPhone 12 is what you should get. I really like the iPhone 11, and I think it's one of the best iPhones Apple has ever made. But the iPhone 12 is better in every way. If you want the best iPhone 11 experience, the iPhone 12 is what you should get.

In this guide, you'll learn the basics of using your iPhone 12 and iPhone 12 Pro Max. We'll discuss things like charging, battery life, storage, notifications, and more!

This book has been updated to reflect the arrival of iOS 15, which will be available for all iPhone models going back to the iPhone 6 in the fall of 2020.

Your gadgets can be easier to use if you read this book.

In this book, you'll find knowledge that's easy to understand, step-by-step, and based on what you need every day. The basics of setting up an iPhone, including data backup and restore, as well as Face ID, email accounts, and screen recording, are covered.

It also includes:

- An overview of the latest features on the iPhone
- The iOS 15 upgrade.
- The new iPhone 12 cameras and voice mail
- iPhone personalization
- As an iPhone user, you have access to Siri.
- Backup and restore for iCloud, iTunes, and Android
- Tricks for the iPhone
- Troubleshooting common iPhone issues

... and lot more.

In this book, aimed at beginners, novices, seniors, and children, you'll find the most up-to-date advice on using your iPhone.

When you're ready to improve your technical knowledge and become the iPhone guru of your dreams, this tutorial is the one you need.

Chapter 1
iPhone Setup

For some, the new iPhone would be vastly different from the previous model, and they would be surprised. In addition, there hasn't been much change in the iPhone setup process. Even if you return to familiar territory, there are still a slew of tiny details to attend to before using your new phone for the first time (or immediately after that).

Here, we'll take a look at how to build your newiphonein the most efficient manner possible.

<u>Setup your iPhone:</u>

If you're using an iPhone that doesn't have Face ID, you won't be able to use Touch ID. When it comes to saving your face, this implies you'll be able to save just one instead of several.

iPhone Setup: The Basics

It's critical that you just re-download the software you'll actually use. This is one of the primary reasons we perform a clean setup on our iPhones, as almost everyone has a plethora of programs installed that we never use. Make use of the App Store and ensure that your Apple accounts are authorized. Using the Updates panel, touch the little symbol to see which accounts you're logged into.) You only need to re-download programs that you've used in the last six months. Take a risk and download files that are regularly on your computer. We're willing to bet it'll be a very small number.

Put up Do not disturb—If you're like most people, your iPhone is constantly inundated with *notifications* and iMessages. This option can be found in the next section down, just below *Notifications* and *Control Center*. ***Set up Do not disturb*** In order to avoid being harassed, you will need to schedule it on a regular basis.

When you want to keep notifications from your face, turn on the alarm and message notifications. When you're able, stay up till 8 a.m. the next morning.

Note: *If you're in a pinch, consider letting certain things through:* Then turn on Repeated Calls by selecting Allow Calls From Favorites and allowing calls from your favorites. iOS 13's **Do Not Disturb** at Bedtime feature, which mutes all alerts and also conceals them from the lock screen, makes it easier to get some shut-eye.

Auto Setup for iPhone

Using Auto Setup on the iPhone 11, you can copy your **Apple ID** and home Wi-Fi settings from another device.

You can simply upgrade to recent version of iOS from an older model of iPhone or iPad if they are currently running previous version of iOS. You can bypass entering your Apple ID and Wi-Fi password by following the prompts. This makes the initial iPhone creation process much more efficient.

Restoring Data from back-up of old iPhone

When you get a new iPhone, you'll frequently be restoring it from a backup of your old one. If that's the case, all you want to do is complete a few things:

- Be sure you include a recent *backup*.
- Consider using Apple's new *Auto Setup* feature to begin.
- The first step is to go to your *iPhone's iCloud* settings and see whether there's a recently available automated back-up option. If you can't find one, then make one yourself. Go to Settings > Your Name > iCloud > iCloud in the menu bar. Take a copy of your data and click **BACK UP NOW** to store it. Wait for it to be finished.

Setup Face ID

It's a lot easier to use and set up **Face ID** than **Touch ID**, but it's not as secure. Simply using the camera, rather than touching the phone with each of your fingerprints, is all it takes. When requested during the initial iPhone setup, do something else in order to activate **Face ID** on your iPhone. Starting from scratch with an already set-up device is possible by going to *Settings > **Face ID & Passcode** and entering your password.*

- Facial recognition is like calibrating your iPhone's compass with the Maps app every so often. Instead of swiping your iPhone from side to side, try turning your attention inward. Two scans are required, and your 3D head might be saved in the iPhone 11's Secure Enclave, which is unavailable to anyone, including iOS itself.
- As with *Touch ID*, in Settings/Settings > *Face ID & Passcode*, you may choose which features to use with *Face ID*.
- If you're a clown, a doctor, or some other type of performer that frequently adopts other characters, you should also adopt a different persona. Make it happen by simply clicking on it in the facial recognition settings.

Create iPhone Email

- You'll need to add your email accounts immediately, regardless of whether you're using Mail, Perspective, or anything like Sparrow. Touch

Settings > Accounts & Passwords, then Add Accounts in the Email app on your iPhone or iPad. To get started, select an email provider and follow the on-screen instructions to enter all of the necessary information.

- You can preview an email before you open it by clicking on the "Preview" button. Isn't it feasible to get as much out of it as possible? The Preview button can be found by going to Settings > Email. If you set your email client to display five lines instead of four, you'll get more information from your emails without ever having to open them.
- Set up your default accounts- Our iOS Email settings may default to a merchant account we've never used, like iCloud, for reasons we've never discovered. Enter your email address, and then click Email in the left-hand navigation pane. When you get to the bottom of the options, you may select the email address you want to appear in new mail by clicking on it. Once there is only one address in this list, you're ready. You can also add other email addresses linked to your email account from there.

Advanced iPhone Email Tweaks

- *Swipe to Regulate Email:* The ability to swipe away emails instead of clicking through and touching on many links is far more convenient when it comes to managing email. It's possible to save a contact to your ***archive*** by swiping in the direction of the *archive*. Alternatively, a *Recycle icon* will appear if your email accounts enable swiping left as a standard ***delete*** action. If you're looking to get through a lot of e-mails quickly, swipe left to *mark* them as *read*. This only affects Apple's built-in email application. Things may be done differently by each third-party email customer.

- Add an *HTML* signature: An ***HTML signature*** can help you appear more professional, so don't forget to include one in your emails. Alternatively, you can copy and paste the code from your desktop into contact and forward it to yourself.

- You can paste it into a contact application by copying and pasting it (or whichever email supplier you prefer, if it facilitates it). Adding a company logo from a web server may be as simple as changing the text in the formatting tags. You'll need an *iOS app* to make one, although they tend to be fairly simple in design.

- Manage *Calendars, iCloud, Communications* and more

- Add default *Calendar* alert: The calendar is great for alerting you to important events, but it's often not at a convenient or beneficial time for you. To ensure that you receive reminders at the right time, set the default timing for three different sorts of events: birthdays, occasions, and all-day occasions. Set up a calendar in Settings > Calendars. Set your Birthday and Occasion reminders to at least one day in advance, as well as your All-Day Occasions on the day of the function by tapping on Default Alert Times (10 a.m.). As a result, you won't have to miss any more events.

- *Background application refresh:* There are a number of apps you'll want to keep running in the background, so check out *Settings > General > Background App Refresh* to see which ones are available. Make sure that *Background App Refresh* is turned on, and then turn off all of the apps that you don't require access to in the background. The best way to find out if you're being slowed down is to turn it off while you're doing something else. Cult of Macintosh Magazine will benefit greatly from having Background Refresh enabled.

Protecting Your Online Activity

- Forms to fill out abound when you're just surfing the web in your browser. There is a good chance that adding your personal information will eat away at your battery life. To ensure that your mobile web browser is built correctly, navigate to *Settings > Browser > AutoFill*. To begin with, turn on the *"Use Contact Information"* toggle switch. Tap on My Info and select the contact you want to use when you see the form fields in the browser. The title and password can also be toggled on in order to maintain that across appointments to the same website. (Because it uses **iCloud Keychain**, you'll need to have access to that as well.)

- To speed things up, turn on the option to turn on your bank cards. Using SSL-encrypted websites is the only way to be sure.

- Use the "***Charge Cards***" button to keep track of which charge cards your iPhone will save you from using. Using the mobile browser, you can add new cards or remove those that you no longer need.

- You may now stop cross-site tracking, which are cookies that allow online businesses to place the same advertising on every subsequent page you visit on iOS 13 and later versions of Safari. The default setting is on, so you don't need to change anything. Take a deep breath and enjoy your newfound privacy.

iCloud is Everywhere:

- In most cases, **iCloud** is everything. There's no doubt in our minds that iCloud is the best and most convenient way to keep your data secure and accessible. Use the *Settings > iCloud menu* to link your **iCloud account** to your personal **Apple ID**. Manage your storage space here, but don't forget to turn on whatever you need right away. Once you've unpacked the iPhone, enable iCloud Drive,

Photos, Connections, Reminders, Browser, Records, News, Wallet, Back-up, Keychain, and more. If you only use Apple's programs and services, you can turn on email and calendars, but you'll likely keep them turned off.

Services Subscription during iPhone Setup

- We recommend turning on *iCloud Photo Library* if you haven't already. Photos and videos are kept safe in the cloud and are available as high-quality backups in the event that you forget the originals. If you have a large number of images in your iCloud Photo Library, you'll want to consider increasing your iCloud storage. To enable the *iCloud Image Library*, go to the *Settings app*, select iCloud, and then select Photos. (It is important to note that this will shut down My Photo Stream.) You'll have to toggle Image Stream back on if you want both.)

- Making use of iTunes Match. If you erase your music files from your iPhone and don't have a backup somewhere, you'll have to rely on Apple Music for whatever quality it delivers you when you hear it. Use iTunes Match to keep your high-resolution music files backed up to the cloud.

- All of your music files are synced or published to iCloud at the highest possible bitrate for your convenience and enjoyment. If your iTunes Match subscription is still active, you can stream or download the music to any device. You'll never have to worry about running out of space on your phone again. Go to the *Settings > Music section*. Once you've done that, you'll be able to sign up for **iTunes** Match on your new iPhone.

More Tweaks to the iPhone's Setup

- Auto-Lock should be extended, because, well, why not? For people who use their iPhones on a regular basis, the two minutes you get by default for the amount of time your iPhone will remain on without turning off its screen is insufficient. You may now set the *Auto-Lock timer* to five minutes under *Settings, General, Auto-Lock*, so you may stop tapping your screen to keep it awake.
- As long as you've set up an **iMessage** account on your Mac or iPad, you'll be able to receive texts

from your iPhone (Settings, Texts, toggle iMessage to ON on any iOS device, Messages Preferences on your own Mac). When you go to Settings on your iPhone, make sure the other device is nearby, and then tap Messages > Text Forwarding. The list will be populated with any and all devices that are made available to it. Turn on your Mac or iPad and see if there is any code on the prospective device. Using the iPhone, type in the code. All of your devices will now receive texts from people who don't use iMessage, as well as iMessages.

- Instead of having difficulty with a Bluetooth speaker, start the EQ in your music app and you will have the capacity to hear your favorite jams. Go to the Settings > Music section. The Clash and Nighttime are the two settings you'll want to use when you want to crank up The Clash while whipping up a quick dinner in the kitchen; this gives you an amazing volume boost.

Chapter 2
iPhone 12 Overview

The iPhone 12 has more new features than virtually any other year in recent memory. As expected, Apple has added a *speedier processor chip* and *improved digital camera*, but these are features we've come to expect. Additionally, *there is an entirely redesigned and reimagined MagSafe charging and accessory ecosystem as well as 5G connectivity.*

Apple and its service providers appear to be aligned in their desire to build this improvement supercycle. Trade-in and payment plans from service providers are extensively marketed, in addition to an increased price of $829 for the basic 64GB model. The market may be difficult because of the economic crisis caused by the epidemic.

For those who are looking for a new phone, it's easy to recommend the default iPhone for the changing times, but it's more difficult to determine whether or not all these brand-new features will lead you to change your phone sooner than expected.

iPhone 12 Pro Overview

In terms of the iPhone, this is a huge year: Apple's iPhone 12 collection is fully redesigned and includes four models, each with a different display size and price point: the **iPhone 12 mini**, the **iPhone 12**, the **iPhone 12 Pro**, and the **iPhone 12 Pro Max**. Across the table, Apple has added *a new MagSafe charging program, a new A14 CPU chip, and all the bluster it can manage about 5G.*

The two 6.1-inch iPhone 12s in the middle of the series are usually very similar, except for the small and maximum iPhone 12s. All of the features of both models, including the **OLED** screen, **CPUs**, and **5G** capabilities, are nearly identical. Additionally, a **LIDAR** sensor, a bit more **RAM**, and double the small storage are all included in the Pro model, which also sports a glossy stainless steel body. All of this might cost you $999, which is about $200 more than the $799 carrier-subsidized starting price of the first iPhone 12.

There are a handful of you who are likely to fork up the extra cash because this is the sparkly one. Due to my maturity and acceptance of who I am, I usually have the same choice. Even so, it's well worth diving in to see if the extra cash is worth it, especially since the iPhone 12 right now comes with an **OLED screen**, which means that differences between the normal iPhone and the **Pro** are fundamentally much smaller than this past year, when the normal model experienced a *lower-resolution* LCD display.

So the real decision for your **iPhone 12 Pro** is whether or not the additional features justify an additional $200 price tag. It may be worth your while to wait a little longer and spend an additional $100 on the **iPhone 12 Pro Max**, which has a larger display and a larger primary camera sensor, as well as an intriguing *new stabilization program* called "*sensor-shift*," which could provide a significant boost in display quality.

As a result, the 12 Pro is in an odd position, and I feel it comes down to how often you plan to utilize the telephoto zoom lens or take nighttime portraits of your family.

Using the iPhone 12's Screen Recording App

What to Know:

- Make sure you put it in the **control center** first. Tap *Settings > Control Center > scroll right down to Screen Recording and tap the + (green plus) logo.*
- Open the *Control Center* by swiping down and tapping the Screen Record icon. After a 3-2 second wait, the recording will commence.
- It's easy to stop recording if you press and hold down the *red status bar* at the top of your screen.

NB: In addition to how to start and stop screen recording, this guide covers how to add the screen recording substitute to the iPhone 12's Control Center.

How to Add Screen Record to Your iPhone 12

To quickly access the controls on your iPhone 12, you'll

first need to add the option to your Handle Center. This is exactly how to do it.

1. Go to **Settings** on your iPhone 12 to begin.
2. Open the **Control Center**.
3. Select **"Screen Recording"** from the drop-down menu.
4. Get as close to the + **(Green plus) logo** as possible.
5. Your **Control Center** has been updated to include the screen recording options.

Screen Recording on iPhone 12

Once you've added the appropriate replacement to **Control Center** on your iPhone 12, you may easily record your display screen.

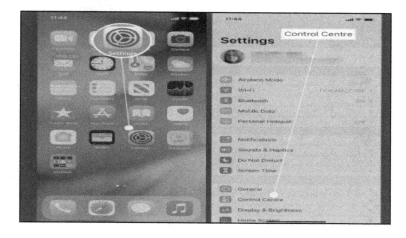

1. Swipe down from the upper-right corner of your iPhone screen.

Tip: *If your iPhone 12 is unlocked, you can do this from the lock screen as well.*

2. Select the "**Display Record**" icon.
3. In order for the saving to begin, you'll have to wait for 3 seconds.
4. Now that everything on your screen is being recorded, you can choose to cease recording.
5. Touch the *red status bar* at the very top left corner of the screen to end screen recording.
6. Select the **Stop** option.

7. The video saves automatically to **Gallery**.

How to Screen Record with Sound on iPhone 12

To begin with, when you take a screenshot, you won't hear a thing. For example, if you want to record your voice as you record the display, all you need to do is modify one parameter.

Swipe down from the upper-right corner of the screen after you **ON** your iPhone.

Tip: *If you're on your iPhone 12's lock screen, you can surely accomplish this.*

1. **Press** and **hold the Screen Record** symbol.
2. As soon as possible, turn the **microphone on.**
3. Press the **record** button.

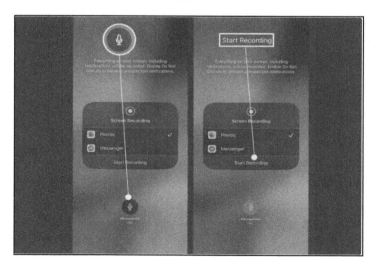

4. You can now interact with the screen by storing it

with sound.
5. To stop recording your screen, tap the **red status bar** at the top left of the screen.
6. Press the "**Stop**" button.
7. The video is automatically saved to **Gallery**.

How to adjust the Settings for Screen Recording

To put it succinctly, no. Rather than recording and saving your photos, the only choice will be the opportunity to accept a Facebook Messenger broadcast. The video's resolution and quality can't be adjusted from the clip.
It's possible to *trim and alter* the web video after saving the display recording in the Photos app.

Limitations For Screen Recording

There are limits to what your iPhone 12 can capture. The problem here is that you can't record loading apps like **Netflix, Disney**, or **Amazon's Prime Movie service**. Due to the fact that pirating the shows you're watching may be in violation of the terms and conditions of your subscription.

The iPhone 12 can generally capture anything, even clips from games you're participating in.

Tip: You may want to adjust the ***Do Not Disturb*** settings after making a display recording so that *notifications and calls* are not captured.

Chapter 3
iPhone 12 and AirPods

AirPods are not included with the *iPhone 12*. There are no headphones or a power adapter included with the *iPhone 12*. There is only a **USB cable** included in this kit. Apple claims to have done away with the earbuds and power adaptor in order to save on packing and weight.

Does the iPhone 12 come with AirPods?

Many people are preparing to upgrade to the iPhone 12 when it goes on sale. It's also a good time to change other important accessories, like your headphones, if you're upgrading your cell phone. *"Do **AirPods** come with the iPhone 12?"* is a common question.

It's a natural inquiry, really. This is a great combination, especially when you're shelling out hundreds of dollars for a new phone and you're also shelling out hundreds of dollars for a new pair of headphones.

We're sorry to disappoint you, but the **iPhone 12** does not include **Air-Pods**. *Air-pods are a separate purchase regardless of which iPhone model you choose, whether* it's the iPhone 12 model or a previous one.

Air-pods, especially the *sound-canceling Air-pods Pro*, are highly recommended because of their great sound quality and impressive capabilities, but they will set you back an additional couple of hundred dollars.

The iPhone 12's Missing Headphones

Adding insult to injury is the fact that Apple has made a significant shift in the components it delivers with new iPhones. Apple's AirPods have replaced wired earphones as the standard accessory for new iPhones. Previously, the charging cable, an electrical adapter for plugging into wall outlets, and wired headphones were included. Not any more.

You only get the charging cable with the iPhone 12 when you buy it. *The Air-pods headsets are no longer included with the iPhone.*

You read it correctly: starting with the iPhone 12, you will not be provided with any earbuds.

Apple claims this decision was made to cut packing waste and shipping weight by reducing the number of boxes needed to ship a product. This will be a part of the company's environmental stewardship.

This is useful in a few ways. It is true that this will lower the iPhone 12's environmental footprint. The Air-pods, on the other hand, will be redundant and possibly wasteful because most people already use headphones.

It also appears to be a marketing ploy by Apple to get people to buy their more expensive Air-Pods. Although **Air-pods** are excellent and worth the purchase price, they do not guarantee that they will be any less expensive.

Advice: Despite the fact that the iPhone 12 does not come with Air-pods, it still has a slew of useful features. We've broken down all of the most significant details about the iPhone 12 into separate sections.

The iPhone 12's headphone options

What are your options if Air-pods and any other earbuds aren't included in the iPhone 12? Everything is possible! Apple still sells Air-Pods for **$19**, which is a good deal. *Air-pods* 2 and *Air-pods* Pro are both available for around **$160** and **$250**, respectively.

However, you are free to choose from a variety of other headsets. Apple's Beats earbuds may be used with the iPhone, as can any other Bluetooth earphones.

The $9 adapter to put a standard headphone jack into Apple's Lightning slot at the bottom of the iPhone *(the iPhone 12 doesn't use USB-C)* is a must-have if you don't want to use Air-Pods. Possibly the iPhone 13 (?).

How to Shut down iPhone 12

What's in it for you?

- In order to see a slider at the top of your screen, you must first press and hold down on the side button, as well as either **Volume Up** or **Volume Down**.
- Finally, move the slider to the very end to **turn it off** altogether.
- Press and hold the side button until the Apple logo design appears on the screen to turn the **iPhone 12** back **ON**.

This little tutorial will show you exactly how to turn off and on your iPhone 12 again. The program also fixes any problems that might happen when you try to turn off your phone.

Even if you have a full battery, you may want to turn off your *iPhone 12* if it's acting up. Restarting an iPhone can fix a wide range of problems. Regardless of the reason, you can turn off the iPhone 12 by following these steps:

1. For a few seconds, hold down the side button and the volume up and down buttons. Let go of the buttons as soon as you see the slider appear at the top of the screen.

Tip: However, holding down the medial side button used to activate Siri on earlier versions no longer works.

2. To turn off the phone, slide it to the right.
3. A progress bar will appear at the center of the screen. After a few seconds, the iPhone shuts down.

Tip: How can you save a lot of battery life on your iPhone 12 by turning it off? There are other options available to you as well. The iPhone's **Low Power Setting** is a good option if you want to save battery life while still being able to use your iPhone. Other ways to extend the life of an iPhone battery are also in the works.

When the Basics Don't Work, Here's How to Shut Down Your iPhone 12

You may have a problem with your iPhone that prevents it from shutting down on its own. If you've followed the instructions above and your iPhone 12 still won't shut down, you may need to try one of these other alternatives.

As a result, a *"force restart"* or *"hard reset"* may be necessary. When the iPhone won't respond to normal methods of shutting it down, this can be used as a kind of restart that clears all of the iPhone's active memory space (don't worry; you won't lose any information like images or communications).

How To Start The iPhone 12

In the end, you'll have to learn how to restart your iPhone 12 if you've successfully shut down your iPhone 12. That can be done by pressing and holding down the side button for a while. Let go of the side switch when you see the Apple logo displayed on the screen. In a matter of seconds, the phone will start up.

How to Restart iPhone 12

Restarting your iPhone 12 is something you'll have to do from time to time. Restarting an iPhone can fix issues like a faulty Wi-Fi connection, an app that has crashed, or other minor issues. If a soft reset is what you need, or if you require a hard reset, it all depends on your specific situation.

Apple's iPhone 12 can be restarted by following these instructions.

- The typical restart turns the **iPhone Off** and turns it **On** again. This is sometimes referred to as a *"normal" reboot*.

If a normal restart fails to fix your problem, or if your iPhone has become stuck and unresponsive, you may need to do a **hard reset.**

Follow these steps to **restart** your iPhone 12:

1. **Press** and **hold** the *volume down* and *side buttons at the same time.*

Tip: To avoid mistakenly taking a screenshot, use the volume-up button instead of the volume-down button.

2. Seconds later, an option to switch off your phone will appear on your screen. Immediately after that,

release both the Side and *Volume Down buttons*.

3. To turn off your phone, slide it to the "**off**" position.

Tip: Have you thought about cleaning the screen on your iPhone? Here's how to do it. I was able to accomplish this without using the phone. this method, you won't inadvertently press anything on the screen or accidentally change your settings.

4. It may take as long as *15 to 30 seconds* for the iPhone to **shut down**. Press the side button once more before the *Apple logo* appears after the iPhone has been turned off for *15–30 seconds*. Allow the *iPhone 12* to restart by releasing the side button.

How to Force-restart Your iPhone 12

Many issues can be resolved with a standard restart, but not all of them. If pressing the side button on your iPhone 12 doesn't do anything, you may need to force restart your device. This is how you do it:

- Press the *Volume Up* key once.
- Press the *Volume Down* key once.

- **Press** and **hold** the *Side button*. The Apple logo will begin to appear as you continue to **hold** the button. Don't worry about the slide to turn off the slider when it's on. As soon as you see the Apple logo, press and hold down the side key until it releases.
- In the meantime, your iPhone 12 will restart.

Chapter 4
iPhone 12 and iPhone 12 Pro Guidelines

Is the *iPhone 12 or iPhone 12 Pro* all you need? You've got the world's fastest smartphone, and by a wide margin. The *A14 Bionic chipset, 5G connectivity,* and a stunning *Super Retina XDR screen* are just a few of the new features included in the new iPhone XR. If you've just gotten your hands on an *iPhone 12 or 12 Pro*, here are some pointers to keep in mind as you get started with it.

1. Move from iPhone

Don't forget to transfer your data first if you're planning on upgrading from an older iPhone. You'll be able to save time by not having to start over with your *iPhone 12 or iPhone 12 Pro*.

While you can do this using **iCloud**, a more expedient method is to use your prior iPhone to transfer your data directly to your computer (if you still have it accessible). Afterwards, connect your iPhone 12 to the same power source as your old iPhone. Your old iPhone's fast boot screen must be visible for this to work. To quickly transfer your data, follow the onscreen instructions and pick Exchange from your iPhone.

2. Familiarize yourself with the meanings of the 5G Icons.

The iPhone 12 finally has 5G connectivity after years of rumors and hoopla. This allows for download and upload speeds of up to *4.0 Gbps and 200 Mbps*, respectively, to be achieved. As a result, three new status icons have appeared: *5G, 5G+, and 5GUW*. Keep an eye out for them.

5G-Indicated that the 5G network is regularly accessible There is a faster high-frequency 5G network available with the designation of "5G+/5GUW."

3. Take control of Smart Data Mode.

5G is a game changer. However, it may also harm the battery life of your iPhone. As a result, your *iPhone 12 or 12 Pro* has the Smart Information Mode feature, which switches between **5G** and **4G** depending on how much data you're using.

As an example, *5G* won't be useful for browsing your Twitter and Facebook feeds. In certain circumstances,

Wise *Data Mode* can help your iPhone take advantage of *4G*. To take advantage of higher 5G speeds, you can download a movie to your iPhone 12 Pro via *Apple TV*.

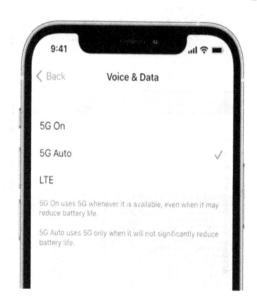

Settings for *Smart Information Settings* may be accessed through the *Settings > Cellular > Voice & Information menu*. Instead of forcing your iPhone to use **5G** on a daily basis, you'll be forced to use it less frequently, which could lead to battery loss.

4. Restrict the use of 5G smart data.

Low Power Mode can even be used to limit the use of 5G on your iPhone 12 or 12 Pro, but only while loading movies. To switch it **ON**, go to Settings > Electric battery.

If you prefer, you can turn on **Low Power Mode** by tapping the icon in the **Control Center**.

5. In the end, don't overlook Magsafe.

You'll find the *MagSafe connector* on your new iPhone. It contains a range of magnets in the rear that allow you to snap on a bunch of add-ons, from cellular chargers to wallets to wise cases. Browse the *MagSafe Add-on* section in the *Apple Store* for plenty of cool items that you should use.

6. Purchase a 20W USB-C power adapter.

You won't find an adapter in your iPhone 12, but you will

find a power to USB-C cable for fast charging. If you're still using a 5-10W charger, you may want to upgrade to a 20W or higher adaptor instead.

7. iOS 5G Updates

Because of 5G connectivity, it is possible to get OTA (over-the-air) iOS updates via mobile data. Go to Settings > Cellular > Information Mode and select Allow More 5G Data to make sure it is possible.

8. HD FaceTime Calls

FaceTime video conversations in 1080p HD are now possible for the first time on an iPhone. Make sure that you've checked the option to allow more information on 5G to FaceTime in HD over 5G in Settings > Cellular > Information Mode. You can even connect to Wi-Fi to achieve this.

9. Improve your low-light photography skills.

The **iPhone 12** and **iPhone 12 Pro** are capable of taking excellent photos in low-light scenarios. The wide camera's larger f/1.6 aperture allows it to capture 27% more light, resulting in images that look far superior to those taken with the *iPhone 11 lineup's equivalent aperture.*

10. Take and Edit 10-little bit HDR Video

With the iPhone 12 Pro, you can capture 10-bit Dolby Eyesight HDR video at 60 frames per second, while the iPhone 12 can only do it at 30 frames per second. You can also use the standard Photos app on your iPhone to edit video clips that you've taken.

11. AirPlay 4K HDR Content

You can **AirPlay 4K HDR** content from *your iPhone 12 or iPhone 12 Pro* to an *Apple TV 2 or an AirPlay 2 compatible smart TV.*

12. Quick-Take Videos

Additional iPhones can be used to download Quick Take. However, it's an extremely useful function that will come in handy when you need to record a brief video. To begin recording video on your *iPhone 12 or iPhone 12 Pro*, simply tap and hold the shutter symbol on the screen to begin. If you want to totally switch to movie recording, simply swipe your palm to the right.

13. Take Advantage of LiDAR

The iPhone 12 Pro has a LiDAR sensor as part of its triple-lens camera array. Many advantages come from using lasers at a step distance. Make sure to check out any augmented reality apps on your new iPhone, for example, so you can get the most out of it. LiDAR also improves auto-focusing by about six times, making it easier for you to take better shots and videos.

14. Snap Portraits in Night Mode

The iPhone 12 and its Pro version enable **Night Mode** on both front and back cameras. You can also take night portraits with the back camera on the *iPhone 12 Pro*. A *LiDAR* sensor's increased auto-focusing enables this capability.

15. Shoot in Apple ProRAW.

A wide, an ultrawide, and a phone lens are all included in your *iPhone 12 Pro's triple 12MP camera system*. Using the Apple A14 Bionic's advanced computational photography functions, it produces stunningly detailed images.

Apple ProRAW, a new image format for the iPhone 12 Pro, is also supported. You can take photos and films in the green format without the use of any fancy camera effects, allowing you to edit and enhance them to your heart's content.

If you purchased your **iPhone 12 Pro** on launch day and have yet to activate **Apple ProRAW**, you will be unable to use it. For the most part, Apple plans to include it in a future **iOS upgrade**.

16. Customize Home Screen

Home display widgets, undesirable apps, and whole pages of the home screen can all be removed or hidden on your **iPhone 12 or iPhone 12 Pro** with **iOS 15,** allowing you to totally customize your iPhone's home screen.

17. Watch Video clips in Picture-in-Picture Mode

Using the *Picture-in-Picture mode*, you can now watch videos on your iPhone or iPad. You should be able to see the movie in a floating PiP pane as soon as you swipe up when watching a full-screen movie. Apps like ***Apple TV, Hulu, and Netflix*** can use the feature. Even in Picture-in-Picture mode, you can view YouTube.

18. YouTube may be seen in 4K resolution.

When it comes to YouTube, the **Google's VP9 codec** is now supported in **iOS 15**. This means that your iPhone 12's stunning Super Retina XDR screen can display YouTube in all of its 4K glory. During a YouTube video, talk about the Quality option and select 2160p to move to 4K resolution.

19. Face Time Eye Contact

Eye Contact can be used in conjunction with **Face Time HD** on the *iPhone 12 and iPhone 12 Pro*. When activated, the feature can give the impression that you are gazing directly at the person on the other end rather than into the camera. To do so, go to *Face Time > Configuration >*

Eye Contact and turn the switch on.

20. Changing the browser's default setting.

You don't have to use the default Mail app or the browser anymore, thanks to **iOS 15**. If you don't like the default browser or e-mail client, you have the option of using any other backed third-party browser (like Chrome).

On your **iPhone 12** or **iPhone 12 Pro**, open the Settings app and select the app you'd want to be the default. You'll then be able to select the app you want to use as your preferred browser or email client.

21. Manage Your Privacy

It's important to monitor your privacy even if you're using one of the greatest smartphones on the market. The good news is that **iOS 15** has many privacy-related settings and tools that can help you accomplish this.

Chapter 5
The iPhone 12 Cameras & Voice Mail

What to know:

- Even the smallest iPhone 12 mini has a better camera than last year's iPhone 11 Pro.
- The iPhone 12 Pro Max's low-light and nighttime photography has vastly improved.

Aside from *MagSafe and 5G*, This year's iPhone's digital camera will be the main reason to buy it. Digital cameras, singular or plural, are another option. I can't get enough of them.

All of last year's Pro models, as well as the **iPhone 12 Pro**, can be found on the smallest, most affordable **iPhone 12 mini**, demonstrating what is possible when super-fast computers are linked to cameras. In spite of this, many *professional photographers* would never use it because it is merely a "*cell phone.*"

An Overview of the Data

Some of the most important features of the **iPhone 12** and **iPhone 12 Pro** are listed below:

- **Night Mode, Dolby Eyesight HDR,** and **Deep Fusion** are now available on all iPhones.
- *Optical image stabilization* is available on all iPhones.
- Both the iPhone 12 and the iPhone 12 mini have a camera that is identical.
- There's a telephoto lens, Apple ProRAW, and Nighttime Portraits on both iPhone Pro models.
- A larger sensor on the primary (wide) camera, a much more powerful telephoto zoom lens, and utilizing the sensor rather than the zoom lens for improved image stabilization are all features of the

iPhone Pro Max.

This is a long list, but perhaps you now have a better understanding of the many functions that are often distributed among the members. Now that we've covered the basics, let's get down to the specifics.

Night Life

The iPhone camera revolutionized itself with the release of the iPhone 11 last year, rivaling and even surpassing standalone cameras in a number of ways.

To capture fine detail in night scenes, one of the most innovative ways was **Night Setting**, which employs image processing to make the images look like they were taken at night (Google's edition can make night time photographs look like they were taken at daytime). All iPhone digital cameras now have this feature, not just the iPhone's wide digital camera.

However, the Pro requires a few things more; the LiDAR digital camera that Apple devotes to the **iPad Pro** is now in the **iPhone 12 Pro**. Self-driving cars can rapidly chart their surroundings thanks to **LiDAR**, which is a **3D** level map of the area that can also be employed at night. The 12 Pro employs this chart to attain practically quick autofocus at night and to enable the background-blurring, family portrait option on Night mode pictures. It's a mind-blowing stunt.

The main (wide) camera on the 12 Pro Max receives a bigger sensor as well. Because bigger sensors have larger pixels, more light can be collected.

ProRAW

The internal supercomputer on the iPhone has processed a photo in trillions of ways before you even see it. In order to create a photograph, a number of photos are combined, the background is blurred, and sensor data is processed. In Apple's brand-new ProRAW format, many of these processes are recorded right along with the photo on the iPhone 12 Pro.

Surely you've figured out how to edit an image in the Photos app and return to it at any moment to make further adjustments. You can put a cool B&W filter system in, then later tweak the blur on the backdrop without affecting other modifications, and you can do that without affecting other edits.

With all of this deep-level running, ProRAW gives the same level of modular customizing as before. It also aids in the storage of sensor-generated "*green*" data. However, the ***ProRAW format*** may also be opened up to designers, so you can do it yourself in the Pictures app. Lightroom will be your editing tool of choice.

How to Configure iPhone 12 Voicemail

What's in it for you?

- Go to your cell phone's voicemail settings and select "Setup Now" from the drop-down menu. There, you'll be able to create a password and save the default greeting.

- Record the custom message by going to Phone > Voicemail, creating a security password, selecting Custom Made, and then recording the message.
- For both options, tap **Play** to hear the greeting and **save** if you are satisfied.

This guide shows you how to set up your voicemail on the iPhone 12 and how to use visual voicemail and other voicemail management techniques.

When you get the iPhone 12, one of the first things you'll want to do is set up your voicemail.

In other words, if you've previously set up voicemail on an iPhone, the process is the same as it was. However, if you've never used an iPhone before, you may find a quick instruction down the page.

1. To begin with, open the iPhone 12's mobile phone app immediately.
2. To access your voicemail, press the voicemail icon on your phone's home screen. Two circles, joined by a right line at the bottom, are seen in this image.
3. A voicemail creation option will appear if this is your first time using the service. To begin the creation process, simply press **Set Up**.
4. Create a voicemail security password when

requested. At least ***4-6 digits*** are recommended for a **strong** password.

Tip: Choosing a password that is easy for you to remember is a good idea. There is no way to recover your password from your iPhone if you've lost or misplaced it. In order to reset it, you will need to contact your service provider.

5. After that, you'll be given the option to choose or create a greeting. Default and custom options are both available.

- **Default:** A pre-recorded message that asks the caller to leave a message by default
- **Custom:** You can create a record that includes the information you want.

When you're ready to save your welcome, select Custom **Tap Record** to get started. When you're finished, hit the **Quit** button. To hear the recorded welcome, press "**Play**" on your phone's voicemail player.

6. When you are done entering your voicemail information and are happy with it, tap **Save**.

What effect does your telephone service provider have on your voicemail preferences?

It used to be that various cell service providers had multiple instructions for setting up voicemail, depending on which carrier you signed up with. Using voicemail on a modern cell phone is a very standard procedure, regardless of the service provider.

Is iPhone Voicemail like Visual Voicemail?

Visual Voicemail is a feature on the iPhone 12 that allows you to see incoming calls. Visible Voicemail will be a voicemail program with a visual interface. Voicemails can be viewed and sorted by you, so you don't have to pay attention to them in the order that they arrived. To avoid hearing messages, you can instead listen to your voice mails in whichever sequence you like.

The iPhone 12's default voicemail is likely to be Visual

Voicemail, as it can be accessed on nearly every mobile service provider network.

Note: Apple has a list of visual voicemail providers in case you want to double-check if you can get the capability from your service provider.

How to Setup iPhone 12's Visual Voicemail Transcription

The majority of US businesses also help with voicemail transcription, which is available on the iPhone 12 in addition to Visual Voicemail. Please follow these instructions to access a transcript of the voicemail on your mobile device:

- Open the phone app on your **iPhone 12** to get started.
- Tap your **voicemail**.
- The software will begin transcribing your voicemails as soon as you tap on one. You should be able to read the transcription on the voicemail page after only a few seconds of waiting.

It is possible that certain words are missing from the transcript because the information is too muddled or confusing for us to properly transcribe.

- After the transcription is finished, you can use the **Share option** to send the voicemail transcription through **AirDrop, email, or iMessage**.

Managing Your iPhone 12 Voicemail

You'll want to change your voicemail password or greeting at some point. Even if the default voicemail notification audio isn't to your liking, you may wish to alter it. All of the choices are straightforward to control.

- Head to your cell phone's voicemail menu and select "**Greeting**." After that, follow the directions to make your greeting more effective.
- **For voicemail security**, go to Settings > Phone and select Switch Voicemail Password. Enter the new password you want to use and click Save Changes.
- Make your voicemail notice sound better by selecting the sound you want from the Alert options in Settings > Sounds & Haptics > New Voicemail.
- Press a voicemail to open it and then tap the Back

option to make a call.

- **To delete a voicemail**, open it by touching it and selecting Delete from the menu that appears. Make sure you don't erase voicemails if you think you might need them in the future, because certain carriers will immediately delete them.

Chapter 6

How to Customize iPhone

Personalize your iPhone text and ringtone sounds.

You don't have to use the identical ringtones and text tones as everyone else to attract attention on your iPhone. If you don't want to open your phone to see who's calling or messaging, you may make all kinds of tweaks, such as altering the sound.

- It is possible to change the default ringtone on your iPhone. Make it easier to be alerted to incoming calls by changing the default ringtone to the one you want. Go to *Settings->Noise (Noise & Haptics on some models)->Ringtone* to change the ring tone.
- In addition, you can set individual ringtones for each of your friends and family members. A love song can play anytime your beloved calls, so you know it's them before you even open your mouth. By pressing on the ringtone you want to improve and then editing it in the phone's Connections menu, you can do this.

- Incoming call screens don't have to be boring when you can get full-screen photos of what's happening. You'll get a full-screen picture of the person who's phoning you if you choose this option. Tap *Mobile phone -> Connections -> tap the individual -> Edit -> Add Picture.*
- Text tones may be customized in the same way that ringtones can be, as can the video that appears when you get a text. Take a look at Configuration->Seems (Noise & Haptics on some models)->Text Tone.

NB: You don't have to use the iPhone's band or text tone. It is possible to buy ringtones from Apple, as well as to use some applications to design your own.

Additional iPhone Customisation Options

Here are a few more options for personalizing your iPhones.

- **Pre-Installed Apps Should Be Removed:** Is there a large number of pre-installed apps on your iPhone that you never use? In most cases, you can just get rid of them! To get rid of an app, use the

67

standard procedures: After you've touched and held them till they tremble, click on the x in the application form symbol.

- The Control Center has a lot more options than you might think at first glance. The tools you use most often can be easily added to the Control Centre. The settings menu may be found under the *Control Center->Customize Settings* section of the Settings menu.

- It's possible to use a third-party keyboard that contains fun features like Google search, emojis, and GIFs in addition to the on-screen keyboard on the iPhone. Make sure you have an app installed from the App Store before you go to *Settings, General, Keyboard,* and *select the keyboard* you'd like to use.

- Is it possible for you to make **Siri** a friend by making her speak in a man's voice? It's possible. Check out **Siri & Search->Siri** *Modulation of Voice->Male in Settings*. If you choose, you may even choose a different accent.

- Make your browser's search engine a different one. Are there any other search engines you'd want to use? Make it the browser's default answer to all of those questions. Make a new selection in *Settings->Browser->Internet Search Engine.*

- As aniphoneor later user, you'll be able to design a wide range of custom gestures and shortcuts for a variety of tasks.

- **This is what happens when you use a jailbroken phone**: The best way to customize your smartphone is to jailbreak it, which removes Apple's control over some personalization options. It's possible that jailbreaking your phone could cause issues with its functionality and security, but it can also allow you more control over your device.

Customize iPhone Home screen

Your iPhone's home screen is likely to be viewed more frequently than any other single screen, so it needs to be set up exactly the way you want it to appear. Here are a few options for personalizing the home screen of your iPhone.

- *Change Your Wallpaper:* You can change the

background image for your apps on the home screen to almost anything. There are a number of alternatives, like a picture of your children or your spouse, or the emblem of your favorite sports team. Go to *Settings->Wallpaper-> Decide* on a New Wallpaper to learn more about the wallpaper options.

- If you're using a live or video wallpaper, do you want to make a statement? Consider using cartoon wallpapers in place of traditional ones. Even though there are many limits, it's a pretty great thing to have. Choose either an active or a live wallpaper from the options in *Settings->Wallpaper->Choose a New Wallpaper.*

- Apps can be grouped together in folders: Use folders to organize your home screen based on the apps you have installed. You should begin by lightly tapping and securing one app until it begins to quiver. After that, simply drag and drop one program onto the other to combine them into a single folder.

- **Add Extra Pages to Apps:** You don't have to put all of your apps on the same home screen. Tap and hold a program or folder, then drag it to the appropriate side of the screen to create *"webpages"* for different types of applications or for different users. See the *"Creating Websites on iPhone"* section of How to Manage Apps on the iPhone Home Screen for further information.

iPhone Customizations that makes things Better to See

There are a lot of adjustments you can make to your iPhone to make it easier for you to view what you're looking at.

- **Try adjusting your screen focus**: if you're having trouble seeing all of the onscreen symbols and text clearly. Screen Move automatically increases the size of your iPhone's display. In order to use this option, go to *Settings->Screen & Brightness->View->Zoomed->Collection.*
- You can increase the iPhone's default font size to make reading easier if the normal size is too small for your eyes. If you want to see larger text, go to *Settings > General > Availability > Larger Text*

71

and turn the slider to On/green.

- It's possible to turn off the iPhone screen's bright colors with the Dark Setting. You may find the basic dark settings in Settings->General->Convenience->Screen Accommodations->Invert Colors.

iPhone Lock Screen Customization:

Like everyone else, you can personalize your iPhone's home screen and lock screen as well. When you pick up your phone, you'll be able to control one of the things you see.

- The iPhone lock screen wallpaper can be changed to an image, animation, or video in the same way that the home screen wallpaper can. For more information, click on the link in the previous section.
- **Using a Longer Passcode:** The longer your passcode is, the more difficult it is to break into your iPhone. Depending on your iOS version, the default passcode is either 4 or 6 characters long. However, you have the option to increase the

length and strength of your passcode. Take a look at *Settings->Face ID (or Touch ID) and Passcode->Change Passcode* and follow the on-screen guidance.

- **Use Siri's Suggestions to Improve Your Life:** Siri may learn about your habits, preferences, and interests, as well as your current location, and then use that information to provide content recommendations for you to read. If you don't like what Siri suggests, you can turn it off by heading to the Settings menu, selecting Siri & Search, and then Siri Recommendations.

Customize Notifications on Your iPhone.

Your iPhone tells you when there are incoming calls, texts, emails, and other types of notifications that you might find interesting. It's possible, however, that these notifications will become grating. Here are some suggestions for customizing your notification settings.

- **Decide on the Notification Style You Desire:** iPhone users have a wide variety of notification options at their disposal, ranging from basic pop-ups to a variety of sounds and texts. The notification choices can be found in *Settings->Notifications->tap the program you want to control->select Alerts, Banner Style, Noises, and more.*

- Is there a way to get multiple notifications from a single program but not have to see each one take up space on my screen? As a notification, you'll be able to bundle all of your alerts into a "stack" that takes up the same amount of space. Go to Settings -> Notifications -> the program form you want to control -> Notification Grouping to set this up on a per-app basis.

- **There is an option for those who do not want to test to get a notification from Adobe:** you can make the camera flash with the Adobe flashlight. A sensitive yet obvious choice can be made in some cases. In *Settings->General->Convenience->Hearing,* turn on the LED Screen for Notifications.

- If your iPhone has Face ID, you can utilize it to keep your notifications safe and secure by receiving notification previews. A simple headline appears in notifications, but when you're through the screen and have Face ID activated, the notice expands to reveal further information about the message. By going to *Settings->Notifications->Show Previews->When Unlocked, you can set this up.*

TIPS: "Reduce Alarm Volume and Keep Screen Shiny with Attention Awareness" is excellent for using Face ID to silence alarms and notification sounds.

- Widgets in the Notification Center can provide additional information. Instead of only collecting your alerts, the **Notification Center** also provides widgets, which are little replicas of full applications that let you accomplish tasks without ever having to open them.

Chapter 7

Siri on iPhone 12

How to Make Use of Siri on iPhone 12

What to Know:

- *In order to use Siri on iPhone 12 models, you have two options:* Siri may be summoned by long-pressing the right-side button or by saying "Hey Siri."
- As an alternative to Siri's previous behavior of taking over all of your screens, it now displays a simple icon and widget reactions.
- Apple products such as *Home Pods* and *Air Pods* can receive messages from Siri in *an intercom-like mode*.

This guide will show you how to activate and use **Siri** on the **iPhone 12**, as well as how to use the intercom feature.

How to Enable Siri on iPhone 12

First, you need to make sure your **iPhone 12** is compatible with the new **Siri** features announced in iOS 15.

1. Open the **Settings menu**.
2. Select **Siri & Search**.
3. Then, on the Siri & Search screen, check to see if the following options are permitted:

- ***Listen to "Hey Siri":*** When you say, "*Hey Siri,*" the voice assistant responds by listening for your command.

- ***Press the Side button for Siri:*** Siri may be awakened by pushing the side button on the right side of the phone for a long time.

- ***Use Siri When Locked:*** This option allows you to use Siri even if your phone is locked.

How to Make use of Siri on iPhone 12

When you activate Siri on your **iPhone 12**, all you have to do is say *"Hey Siri"* or press the key on the right side of the phone for a few seconds.

Siri's full-screen capability has been removed with **iOS 15's Siri upgrade**. To indicate that Siri is listening, you will see a bright indicator at the bottom of the screen. Replies can then appear as widgets or banners on parts of your mobile phone's screen, but they won't take over the entire display.

iOS 15 brings Siri some much-needed Updates.

A number of enhancements were made to Siri in **iOS 15** (the version that was pre-installed on iPhone 12s at the time of release). Additional or enhanced functions have been added or improved in addition to the improvements in appearance that were previously stated.

- ***Better Answer:*** Apple's Siri handles more than 25 billion requests per month for *Apple*. All those responsibilities have required the *VA* to learn a great deal. According to Apple, there are now 20 times as many facts available to Siri as there were a few years ago. In addition, the digital assistant's internet-based solution-providing capabilities have been improved.

- ***Insightful Recommendations Shortcut Suggestions***, a new Siri widget, make it easier for Siri to recommend actions you do on a regular basis. As an example, when you get into your car, Siri might suggest that you open up Google Maps or buy a cup of coffee from your favorite coffee shop. Also, you can put these ideas on your home screen.

- ***Sharing an estimated time of arrival:*** When you're on the phone with another iPhone user, ask Siri to "Talk about my ETA."In comparison to that person, Siri will be able to offer your **ETA** via Apple Maps. However, there are always a few things to keep in mind. There is a strong likelihood that your **Apple ID** email will be used to communicate with you about your expected introduction time.

- ***Voice Messaging:*** For example, Siri can record and transmit sound messages over **iMessage** or even **MMS texts**. Voice Messaging (as a result, you can send audio recordings to Android users).When you tell **Siri** to *"Send an audio message to,"* she will record and send the audio. Before you submit the information, you should be able to hear it, cancel it, or rerecord it. Alternatively, you can use **Car Play** to accomplish this.

- **Improved Translation:** With **iOS 15**, Siri's

translation capabilities have been improved. There are *65 language sets* available now, and you don't even have to be connected to the internet to use them. There is also a noticeable improvement in the translations.

- ***Cycling Directions With Maps***: You may now ask Siri, *"What exactly are the cycling routes to [name of destination or area]?"* "(In the event that Siri doesn't know or recognize the location you requested directions for, the tone of the voice assistant will provide you with ideas and refer you to Apple Maps.
- **E-mail reminders**: Siri now has the ability to recognise possible reminders from e-mail and create suggestions in iOS 15. Using this feature can help you stay on top of your game if you find yourself getting a lot of reminders from your e-mail.

How to Make use of Siri as an Intercom

When using Siri on the iPhone 12, you may also want to take advantage of the intercom feature if you have other Apple devices in your house. Apple products, including

iPhones, iPads, Apple Watches, AirPods, and CarPlay, all have an intercom feature.

Hello Siri, please tell everyone [your information]. If you're using Siri, you'll be able to respond to the messages by saying *"Hey Siri, replay [your reaction]."*

HomePod and AirPod devices will be able to display notifications for messages made using the intercom feature, which will be a pleasant addition to the user experience.

Chapter 8

How to Take Screenshot on iPhone 12

What to Know:

- To take a screenshot, press the volume up button and the side button simultaneously.
- In the Pictures app, in the Screenshots area, most screenshots are saved for later use.
- To share the screenshot, open up the Pictures app, tap the screenshot, then hit Talk About > Select an app from the drop-down menu that appears.

How to take a screenshot on an *iPhone 12* and how to share it are all covered in this section.

Take a snapshot on your *iPhone 12* to preserve vital information, a funny story, or any other memorable moment. However, there are third-party apps that allow you to take screenshots. iOS now has the option to take a screenshot on the *iPhone 12*. Here are a few pointers on what you should do:

1. Using your iPhone's camera, take a screenshot of anything you want. This could be a piece of literature, a web page, or even an app.
2. Push the volume up and side buttons at the same

time to activate the side button.
3. A screenshot is taken when the screen flashes and a camera sound is heard. The screenshot's thumbnail appears to be in the screenshot's lower-left corner.
4. Swipe the screenshot from the edge of the screen to save it and do not do anything else with it. Simply tap the thumbnail to make changes or do other actions on the screenshot.

How to Find Your iPhone 12 Screenshot

In your phone's pre-installed Pictures app, you can save screenshots of **iPhone 12** to a specific folder. Follow these steps to get a screenshot:

1. Tap the **Pictures app**.
2. **Albums** must be selected in the bottom bar if they are not already there.
3. **Scroll down** and **tap Screenshots**. Every screenshot you've ever taken is here in one convenient location.
4. You can also find screenshots in your Camera album, along with other images.

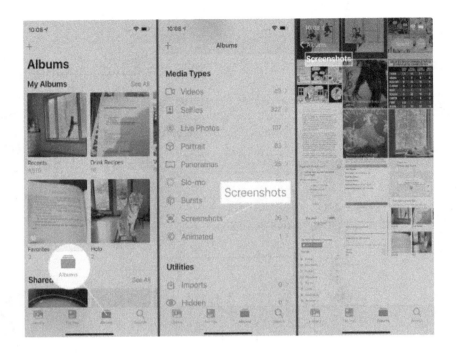

How to Share iPhone 12 Screenshots

As with any other image, you can share a screenshot from your iPhone 12 by sending it as *a text message, email, posting it to social media, etc.* You have the option of deleting or syncing it with your computer. To send a screenshot, use one of the following methods:

1. The **screenshot** can be found in either the *Camera Album* or *the Screenshots Album* in the Pictures app. To view the image, first tap on the screenshot.

2. Tap the "**Share**" option.

3. Select the app from which you want to send the screenshot by tapping on it.

4. The following row of apps will launch when you tap on them. App-specific sharing options should be utilized to their fullest extent.

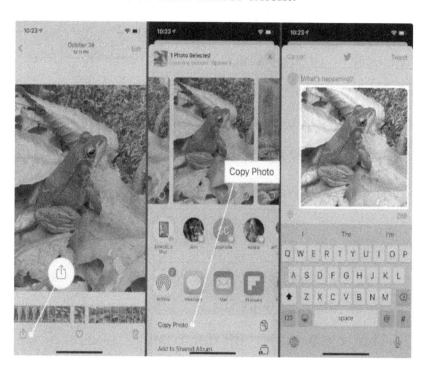

How to Close Apps on iPhone 12

What You Need to Know:

- To close an app, swipe up from the display's bottom, slide left, then swipe up and off the top of the screen.

- By swiping many more fingers at the same time, multiple apps can be shut down simultaneously.
- There is no built-in option for deleting all apps at once.

How to close apps on the iPhone 12 is described in detail in the information provided below. It also busts the idea that getting rid of apps saves a lot of battery life.

Halting apps, force stopping apps, or force closing apps are all synonyms for *shutting off apps*.

Using these instructions, you can close apps on your iPhone 12:

1. Swipe upward from the bottom of any screen on the iPhone 12 (the home screen or an app). It's possible to swipe as far as you like, but a quarter of the way up is plenty.
2. This exposes all of the running apps on your iPhone 12's screen.
3. To see all of the apps, simply swipe back and forth.
4. Swipe up and off the top of the screen when you find the one you want to exit. The app will shut down if they disappear from your screen.

Tip: *You can close multiple apps at once. Swipe them all at once with a few more fingers.*

5. You can only close up to three apps at a time on the iPhone 12. There is no built-in way to close all of the apps at once.

When You Should Close iPhone Apps

When you don't use an iPhone app, it goes into the background and freezes. As a result, it consumes very minimal battery life and is highly unlikely to access any personal data. In most circumstances, an app that freezes is the same as one that has been closed. Frozen apps restart faster than apps that have been closed when you open them.

Close Non-Working Apps

The only time it's necessary to shut down or exit an iPhone app is if it's not working properly. If this is the case, a quick restart of the app could alleviate a short-term problem, much like restarting your iPhone can.

The sole purpose of some apps may be to ask the computer to allow them a specific amount of time and power to complete a task or even to continue operating (think of music, mapping, and marketing communications apps).

Chapter 9
iTunes, iCloud & Android Backup Restore

To activate your new iPhone device, you don't need to use your computer or a wired internet connection. All you need is a mobile data connection. In order to return to the previous screen after completing the setup process, you can tap the trunk arrow in the upper left corner of the screen and then scroll to the upper right to access another display.

You can begin by tapping the power button on the top edge of the newiphone. If you see a vibration while holding theiphone down, that means the device is starting up.

Once the computer has finally started up, you can proceed with the following steps:

- Start the wizard by swiping your finger across the *screen*.
- It's easy to find *English* because it's often near the top of the list of available languages. However, if you want to use a *different language*, simply scroll down to the bottom of the page and touch on the language of your choice.
- Select your nation of origin, such as the United States, which may be near the top of the list. Select "*America*" or all of your choices if necessary; otherwise, move on down the list.
- An online connection is required to activate your *iPhone*. If you have a *Wi-Fi connection*, you can try this. To select a *specific network*, look through the list and touch on the name of the network that interests you.
- After entering the **Wi-Fi password** (*usually found*

inscribed on your router and known as the *WPA Key, WEP Key, or Password)*, select Subscribe. As soon as you see the tick, you know you're connected, and a radio picture appears near the screen's surface. Apple will now activate the iPhone by default. The process could take some time.

- Check for *updated internet Settings* after inserting a new sim card if you have aniphone with **4G**. Since you can try this at any time, tap Continue to continue for the time being.

- Location services help you with things like maps, weather apps, and more by giving you precise information about where you are. By accepting location services, you can decide whether or not to make use of location-based services.

- **Touch ID**, Apple's fingerprint identification system, will now be required. By using *Touch ID*, you can use your fingerprint instead of a passcode to unlock your new *iPhone*. Put your finger or thumb on the home button (but don't press it down!) to activate Tap **Identification**. Tap to install Touch ID later if you want to skip this step for the time being.

- You'll be guided through the process of setting up *Touch ID* by the instructions that appear around the screen. Once theiphone has properly scanned your fingerprint, remove your finger from the home button. Upon completion of the scan, you will get a screen informing you whether or not the tap recognition was successful. Choose Continue.

- To keep your *iPhone* safe, you'll have to enter a passcode. If your fingerprint isn't recognized when you use *Touch ID,* you'll need to enter a passcode. The iphone gives you many options for safeguarding your computer data. Select your lock mechanism by tapping the password substitute.

- It's possible to set up an alphabetic and numeric code (a password that includes characters and figures) as well as a custom alpha-numeric code. or an alphanumeric code of four digits. It's possible that if you haven't installed or created *Touch ID,* you can choose not to enter a password for your account. Tap on the *security option* that best fits your needs.

- For added protection, I recommend setting up a 4-

digit number code or *Touch ID*, but all other options may be set up in the same way. Using the keyboard, type in the password you've chosen for your account.

- Re-enter your *account password* to be sure it's correct. You'll be prompted to try again if the password doesn't match. You'll immediately move on to the next display if they match.

Your *iCloud or iTunes backups* can be utilized to restore all of your apps and data on your new phone during the initial set-up process if you've previously used an iPhone. However, if you're switching from Android to an iPhone, you can choose to transfer all of your data by making a decision and then making the choice you desire.

How to Move Data From an Android Phone

Your computer data may be easily transferred from an Android OS device to your newiphone.

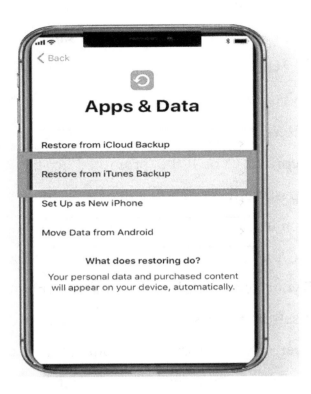

Access the **iOS** app now. Use the application! I'll show you how to transfer your info!

- Tap to migrate data from Google Android on the setup wizard screen for the iphone.
- Use the Play Store on your Android device to download the app recommended by the setup process and install it. When the app is installed, open it and select Continue to see the Terms & Conditions.
- To begin linking your devices, press **Next** on your

Android device. Select the **Continue** option on your own iPhone.

- While using an Android device, you'll see a *6-digit code* appear on the screen of youriphone.

- All of the data that's being transported would be screened by your Android device. All options are checked by default, so tap the related collection to deselect anything you don't want to see. In order to proceed, simply tap Next on your Google Android phone or tablet to proceed.

- As the change progresses, theiphone display screen changes to show the position of the data transfer as well as a progress report.

- A confirmation screen will appear on each device when the transfer is complete. To exit the app, press ***Done*** on your Android device, then tap the Continue button.

- Having an *Apple ID* on your iPhone is vital since it allows you to download apps that are compatible with your device and synchronize data across numerous devices. Your Apple ID should already exist if you've previously had an iPhone X or downloaded music to your laptop using iTunes.

Log in with your e-mail address and password (when you have lost or forgotten your **Apple ID** or password, you will see a link that may help you reset it). A free **Apple ID** is required if you're a first-time user of Apple's newest devices.

- If you have an iPhone, you can view the terms and conditions of use. Please take a look at them (you can touch on more to learn more) before you tap Agree.

- You'll be prompted to choose whether or not you want to sync your data to **iCloud**. This is to make sure that your otheriphone data, including bookmarks, connections, and other items of data, is supported securely. If you don't want to combine, you can do so by pressing the merge button or by not merging if you want to store your information elsewhere.

- For online and app purchases, you can utilize Apple Pay, the company's secure payment system, which maintains your encrypted credit or debit card information on youriphone as well as your fingerprint. To proceed, click **Next**.

- In order to feature or add a card, place it on a flat,

stable surface and then place the iPhone on top of it. When you use a credit card, the information is automatically scanned, and you are prompted to double-check that the information displayed matches the information on your card. On the back of the card, you'll also be asked to enter the card's **CVV** (safety code). By tapping the links, you can enter your credit card information manually if the camera is unable to recognize your card. To avoid setting up *Apple Pay*, you can just tap the *Create Later option*.

- Apple's secure method of sharing passwords and payment information across all of your Apple devices, is discussed on a separate screen in the app. If this is your first Apple device, you may be asked to enter your *iCloud security code* to verify the device and import your current data, or you may be required to register your keychain again. If you don't want your credentials to be shared with other devices, don't use the *iCloud keychain* or remove them from the device.

- Setting up your **Apple keychain** requires you to enter a security password (the same one you used on your *iPhone* or generate a new code that's displayed to you during the setup process. Theiphone will prompt you for your *iCloud* security code, which you should enter.

- When logging on to an *iCloud* safety code, you'll receive a verification code through SMS. A text link to your smartphone text code may be useful (if you haven't already done so with Apple) in order to make the code more easily accessible. On your iPhone, enter this **code** if necessary, and then select Next to proceed.

- **Siri** is the next thing you'll be asked to build. In other words, **Siri** is your own personal digital assistant who can do everything from search the internet to send messages to check your device's data, all without the need to open any specific apps. If you'd prefer to activate **Siri** later, you can do so by choosing the option instead of creating **Siri** right away.

- The iPhone requires you to say a number of sentences to review your speech patterns and recognize your voice in order to establish and develop **SIRI**.

- There would be a tick at the end of each term, indicating that the listener had understood it. *Alternatively, you may be advised to read aloud.*

- When the five words have been completed, you will get a message on the screen indicating that Siri has been successfully set up. Choose **Continue**.

- The display of the iPhone changes the color balance in order to make the screen appear more natural in different lighting scenarios. After theiphone has finished configuring it, you can turn it off in the screen settings. To proceed with the setup, use the "**Continue**" button.

- You should check if your iPhone has been reloaded. To begin the process of transferring data from your old iPhone to your new iPhone, tap the Start button.

- After downloading apps and information, you'll be asked to check the battery life of your newly purchased iPhone. To confirm, press the **OK** button.

- You'd get a notification on your apps if they were downloading in the background.

NB: In order to set up any new iPhone: There is a similar procedure to the one mentioned above.

iPhone Backup Restore from iCloud or iTunes

This will allow you to restore your iPhone from a backed-up iTunes file if you have iCloud and the most recent version of iTunes installed on your computer. On your own iPhone, tap restore from iTunes back-up and connect it to your personal computer. The laptop's screen will display instructions on how to recover your computer's data.

If your old iPhone model was supported by **iCloud**, then you can restore your apps and data to your new device by following these steps:

- Select the option to restore from iCloud.
- Enter your old iPhone's **Apple ID and password** when signing up for a new account. A link on the website can help you reset your account password if you've forgotten it.
- The Terms & Conditions page would be displayed. Use the links to learn more about specific topics. Select "Agree" if you're ready to move forward.
- Creating an **Apple ID** and connecting to the iCloud server will take some time on youriphone.
- You'd see a list of backups that were available for download. The most recent backup is likely to be at the top of the list, with a second option below. You can examine the available options for restoring from a backup by tapping on the screen for all of the backups.
- To begin installation, tap on the *backup* you wish to *restore*.
- You'll be able to see the status of the download in the form of a progress bar. These devices will

resume after the restore process is complete.

- Your *iPhone* will receive a notification when it has been successfully upgraded. Choose *Continue*.

- Your iCloud (**Apple ID**) password must be re-entered to complete the iCloud setup on your recently recoverediphone. Tap Next after you've finished entering or reviewing it.

- The security information linked to your Apple ID will prompt you to upgrade. Tap on any stage to remove or bypass this application and all of its data. It's okay if you're unwilling to do so; simply press a different button instead.

- On the iPhone, Apple Pay is Apple's secure payment system that keeps encrypted credit or debit card data on your device as well as uses your fingerprint to make safe transactions with other apps. To continue, click **Next**.

- A card can be added to an iPhone device by sticking it to a flat surface and then placing the iPhone device on top of it. This allows the card to become part of the camera system. Automatic scanning of your credit card information will take place, and you will be asked to confirm that the information displayed matches what is on your card. On the back of the card, you will also be required to enter the **CVV** *(safety code)*. If you like, you can enter your credit card information yourself if you like (*or the camera will not detect your cards*). To avoid setting up Apple Pay, you can just tap the *Create Later option*.

- The **iCloud** keychain, Apple's secure means of transferring your stored account and payment information across all of your Apple devices, is also discussed on this screen. As long as you have an **Apple ID**, you can use the iCloud security code to authenticate and import your current data from your old device. Don't use *iCloud Keychain or restore passwords* if you don't need to share important info with other devices.

- Your iPhone account password (*the same one you'd establish on your own iPhone*) would be requested if you chose to create your *Apple keychain*. When prompted, wear your **iCloud** security code on youriphone if you're comfortable with it.

- When logging into an iCloud safety code, a

verification code will be sent to your mobile phone through SMS. It is possible to hyperlink your smartphone text code so that it can be delivered as a text message (if you have never distributed one with Apple before). On your iPhone, enter this code if necessary, and then select Next to proceed.

- Next, you'll be asked to create an instance of Siri. Using Siri is like having a personal assistant who can search the web for you, send messages, examine your device's internal data, and more. Tap the decision to start Siri or wait to start Siri later if you want to skip this for the time being.
- To set up and establish **SIRI**, you'll need to say a lot of words to theiphone in order to learn your conversational patterns and recognise your voice.
- A tick will appear after you've said each term, indicating that you've understood it. Reading aloud may be indicated by another expression.
- Following the completion of the five sentences, you'll be greeted with an alert informing you that Siri has been appropriately programmed. Choose **Continue**.
- To make the screen appear more natural in

different lighting conditions, the color balance of theiphone display has been altered. After theiphone has finished configuring it, you can turn this off in the screen settings. To continue utilizing the setup, press the continue button continuously.

- *What's the status of your iPhone?* To begin transferring data from your computer to your new iPhone, simply *press* the "**Start**" button.
- Make sure your newiphone has enough charge to keep these devices from shutting down when downloading programs and information. To confirm, press the **OK** button.
- In the background, you'd get a *notification* about the programs you'd like to download.

Please note: *Setting up a new iPhone model requires a similar procedure to the one mentioned above.*

Chapter 10

How to Restart an iPhone

The iPhone is a powerful computer that can be carried about in the palm of your hand. Like a computer, an iPhone may need to be rebooted or reset to fix an issue. Restarting an iPhone is as simple as turning it off, then back on again. Attempt a factory reset if your iPhone won't restart. Neither of these methods removes information or settings from the iPhone. A restoration erases all of the iPhone's content and returns it to factory settings, and you may also recover your computer data from a back-up .

How to Restart the New iPhone

In order to fix common issues such as poor cellular or Wi-Fi connectivity, program crashes, and other nagging hiccups, you may need to restart your iPhone. Apple has assigned additional functions to the sleep/wake button on certain smartphones. To activate Siri, discuss the emergency SOS feature, or for any other purpose. As a result of this modification, the restart procedure differs from prior models' techniques.

To restart an iPhone 12, iPhone 11, iPhone X, and iPhone 8:

- At the same moment, press and hold down the Rest/Wake and Volume Down buttons. Increasing the volume does the trick, although it's possible to accidentally take a screenshot while doing so.
- Let go of the rest/wake and volume down buttons when the Glide to Power Off slider reaches the top.
- Turning off the phone is as simple as moving the slider to the left.

How to Restart Different Models of iPhones

Other iPhone models can be restarted in the same way that the iPhone can be turned on and off. The following are the objectives to be met:

- The sleep/wake button is located at the top of the phone on older models. From the iPhone 6s and beyond, it's on the right side of the phone.
- Release the Rest/Wake button when the screen's

energy off slider shows.
- Turning the energy off slider to the right causes the iPhone to shut down more slowly. The shutdown is in progress, as indicated by the presence of a spinning icon on the screen. It's possible that finding out will be difficult in the poor light.
- Press and hold the Rest/Wake button until the phone turns back on.
- Release the Sleep/Wake button when the Apple logo appears over the screen and wait for the iPhone to finish rebooting.

How to perform a hard reset on the iPhone.

The basic restart fixes many issues, but it doesn't always answer the majority of the questions. It's necessary to use a **"*hard reset*"** option when the phone freezes up and won't respond to tapping the Rest/Wake button.

The hard reset method on the *iPhone 12, iPhone 11, iPhone X, and iPhone 8 series* is different than on earlier devices. To perform a hard reset on these iPhones, follow these instructions:

- Click and release the *Volume Up* button.
- Click and release the *Volume down* button.
- Press and hold the *Sleep/Wake button* before gliding to power off slider appears.
- To reset the phone, slide the force-off slider all the way to the left.

Other iPhone Models: How to Perform a Hard Reset.

Restarting the phone and reloading the application storage are two benefits of performing a hard reset. In most cases, it aids in a fresh start for the iPhone instead of wiping everything. Usually, a hard reset isn't necessary, but if it is on an adult model (except for the iPhone 7):

- To turn off the phone, press the *Sleep/Wake* and Home buttons simultaneously while it is on the phone's screen facing you.
- Keep your support for the links going. Never release the links when the energy off slider is all the way up.

- Let go of the sleep/wake and home buttons when the Apple logo appears.
- Wait for the iPhone to reset.

How to Perform a Hard Reset on an iPhone 7 series

For the *iPhone 7 series*, the hard reset method is a little different. As a result of the home button being a 3D Touch-panel rather than a physical button, this is the case. As a result, Apple redesigned the process for factory resetting these models.

If you're using an *iPhone 7 or 7 Plus*, keep the Volume Down and *Sleep/Wake buttons* pushed together at all times.

More Help Resetting Your iPhone

When an iPhone has a problem that is too hard to fix with a restart or reset, it may need to be replaced. To fix the problem, follow these advanced troubleshooting techniques:

- **Stuck at Apple Logo**: If your iPhone gets stuck at the Apple logo when it first starts up, a simple restart will not fix the issue. Take the phone to an Apple store or a service center for repair.
- **Restore to default Settings**: Restoring an iPhone to its factory default settings can fix a few tricky issues, such as erasing all of the data on the device and starting over from scratch. Before you put your iPhone up for sale, make sure it's in factory mode.
- **Recovery Setting** Try recovery mode if your iPhone is stuck in a reboot loop or you are unable to see the Apple logo at startup.
- **DFU Mode**: DFU (Disk Firmware Update) mode is used while downgrading or jailbreaking the phone.

Chapter 11

How to Show Battery Percentage on an iPhone 12

What to Do:

- Swipe down from the top right corner of the screen.
- Before the icons start wiggling, **tap** and **hold** the screen's battery area in the upper right corner, close to the battery symbol. Add a widget by pressing + > *Battery* > *Widget design* > *Include Widget* > *Done*.

In addition to showing the battery percentage as a widget on your home screen, the information provided here shows you how to do both on *the iPhone 12*.

Previously, you'd have to switch **ON** the battery percentage replacement in order to get these statistics on previous versions of **iOS**. Not even the new *iPhone 12* has swayed me! By default, the electric battery percentage feature is now enabled. However, you must know where to find it.

Notice: *that the battery indicator and percentage cannot be shown simultaneously if you are upgrading from an* *iPhone without Face ID because of the camera notch at its very top.*

1. Swipe down from the right side of the *iPhone 12's display to open Control Center.*
2. The battery percentage may be found in the upper right corner of the screen, directly next to the battery icon. This is a way to extend the life of the battery on your iPhone 12.

3. The **Control Center** can be closed by swiping up or even tapping the backdrop.

The only thing you need to do is check the battery's capacity on a regular basis. Consider adding a battery status widget to your home screen if you want to keep an eye on how much juice is left on your device.

Tip: You can ask Siri to tell you how much juice is left in your iPhone 12's battery. ***Siri may be activated by pressing the Siri button.*** By pressing the side button, ask Siri, *"Hey Siri, how much battery life do I have left?"* The battery level may be shown on the display.

Adding a Battery Widget to Your iPhone 12

A battery percentage widget can be added to your home screen with iOS 15, which is pre-installed on the iPhone 12. This is how you do it:

1. Before the icons begin to *wiggle*, **press** and **hold** the screen.
2. Press "+"
3. When a new widget pop-up window appears, select ***Batteries***.
4. Select the widget design that you want to utilize. To see the options, swipe back and forth. If you have an ***Apple Watch or AirPods*** connected to your phone, the Batteries widget may also show information about the battery pack.

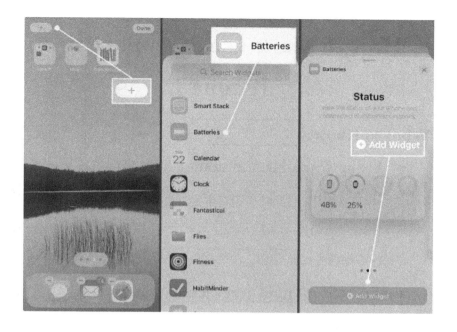

5. Click on the **"Add Widget"** button for the widget you want to utilize the most.
6. In the second step, the widget is installed on the home screen. Tap **Done** after you've got it where you want it.

How to Make use of Apple Pay on iPhone 12

What to Know:

- Using **Face ID** and double-pressing the switch on the right side of the phone, authorize the transaction and keep your phone close to the payment terminal.
- To add a new transaction card to your Apple Budget, tap the Add button. Cards linked to your Apple accounts may already be included in your Apple Budget.

Instructions on how to set up **Apple Pay** on an **iPhone 12** and use it to make payments at **NFC terminals** are provided in this section.

As a convenient feature of contemporary iPhones, **Apple Pay** is present on the iPhone 12 as well. In case you've never done so before, here's an overview of the process of setting it up.

1. To find **Apple Pay**, tap *Resources > Wallet*.
2. On the display, there is a summary of how Apple Pay works. Take a look at it and then press the

Continue button.

3. Select an existing credit or debit card or enter a new one by tapping **Credit** or **Debit Card**.
4. The cards linked to your *Apple account* will show up on a separate screen. If you intend to use a specific credit card, go with that one.
5. Tap the **Add a Different Card** button to add a new card.

Tip: *You may only be able to add a new card to your account if you have no other cards linked to it.*

6. Afterwards, you will be asked to scan a card. Following completion of this step, take the next step by pressing the **Continue** button.

Tip: *If your card isn't being accepted, you can try manually entering the details.*

7. You'll be prompted to enter the card's security code. Then press the **Continue** button.
8. You'll see a confirmation screen when more cards are added. Take the next step by pressing the **Continue** button.

Note: Please be aware that you may be asked to learn about and agree to the terms and conditions. If this is the case, read the provided details and then tap **Agree**. You won't be able to add your transaction card if you select **Disagree**.

9. A new help screen displays, this time explaining how to make payments with **Apple Pay**. To return to your **wallet**, go through it and then tap **Continue**.

How to Make use of Apple Pay in-store

You can utilize Apple Payout in participating stores once you have enrolled a minimum of one credit or debit card in your **Apple Wallet**. In order to be approved by **Apple Pay**, this is all it does.

In the event that you encounter a certain sign, simply follow these steps:

1. In order to activate this feature, you must press and hold the right-side medial key on your iPhone 12.
2. You can use **Apple Pay** with your default credit card. **Face ID** can be used to verify your identity while you're still holding your phone.

Note: Even if you have numerous credit cards in your wallet, you can still use a different one. When your default card displays, touch it and choose the credit card you intend to use.

3. Hold your phone near the payment terminal and you'll see "**Done**" and a blue checkmark appear on your screen.

How to Modify Your Default credit Card on Apple Pay on iPhone 12

If you only have one credit card linked to Apple Payout, that card will be used for any future purchases. If you add or change credit cards, you may want to replace the default card with a different one.

Simply open the Wallet and tap and hold the card you want to use as your default to make the process quick and straightforward. Afterwards, move that credit card to the first place in your list of cards. That way, it'll always be set to "**default**."

For those who are struggling, here's an alternative method for setting a different card as the default instead:

1. Start by going to the **Settings app**.
2. Go to the bottom of the page and select **Finances & Apple Pay.**

3. From the list that appears, choose your **default credit card**.

4. Select the **new credit card** that you plan to use as your default card.

Nest time, As soon as you double-press the *side button* to activate **Apple Pay,** the card you selected as the new default will appear.

Chapter 12
iOS 15

Apple unveiled *iOS 15,* the most recent version of its *mobile operating system*, in June of 2020. This is one of Apple's most recent iOS upgrades, bringing changes to the **Home display**, new features, updates for existing apps, and improvements to Siri, as well as a slew of other improvements to the iOS user experience.

First and foremost, *iOS 15* introduces a redesigned Display Screen that includes support for widgets for the first time. It's possible to drag widgets from your watch to the home screen and pin them in various sizes.

Widgets can be surfaced by the iPhone based on time, location, and activity using the Stack function. Widgets for *business, travel, sports, and more* can be added to any home screen page. There is a widget gallery where consumers may choose new widgets from apps and personalize them. The space where widgets are generally housed has also been updated.

After completing the app pages on an iPhone, you'll be taken to the new App Library, where you can see all of your iPhone's apps in one place. In addition to the built-in folder system, Apple has established folders like *Recommendations and Apple Arcade* that automatically surface area apps. To keep your *home scree*n clutter-free, new app downloads can either be added to the home screen or stored in the *App Library.*

Incoming calls and *Siri* requests no longer fill up the entire screen thanks to new space-saving measures. Activating *Siri* displays a small cartoon *Siri* icon at the bottom of the display screen when calls *(including Face Time/VoIP phone calls)* come in via a small banner on the iPhone's screen.

FaceTime or a video can be viewed in a small window that can be resized and relocated to any part of the iPhone's screen while the user is using other apps at the same time with an image in *Picture mode.*

On iOS 15, **Siri** is more intelligent and can answer a wider range of questions using information retrieved from the web, and Siri can also deliver audible messages. Dictation through a device's keypad includes an extra layer of privacy protection for the dictated texts.

Apple introduced App Clips in iOS 15, allowing users to benefit from specific features of an app without having to download the complete application. Checking a code rather than downloading an entire app can allow you to accomplish things like renting an electric scooter, buying a cup of coffee, or even filling up a parking meter without having to download an entire program.

Apple describes App Clips as merely a *"mini section of a decent app face"* that is only meant to be used when it is most convenient. *Apple-designed App Clip rules, NFC tags, or QR codes* can also be discussed in Communications or Browser.

To keep important conversations at the top of the text messaging app, Apple now allows users to **"pin"** conversations. To pin any communication, simply swipe to the right on any. It is now possible to respond directly to a specific message in a conversation, which is particularly useful for team chats.

When a user's name is mentioned in a group chat, Apple has included a **"@mention"** function, which allows the chat to be muted but still sends a notification. Adding a photo or emoji to the team chat photo can let you know who was the last person to speak, while the icons at the top of each conversation show who was speaking most recently.

In addition to new haircuts, headgear, face coverings, and ages, new **Emoji** *options include emoji stickers for an embrace, a fist bump, and a blush.* Because of the new facial and muscle foundation, **emoji** are more expressive than before.

To allow children to use an Apple Watch without an iPhone, iOS 15 and watchOS 7 work together to make it possible for parents to set up and manage mobile Apple Watches for their children through Family Setup.

To help users better understand how audio amounts affect hearing health, Apple has included support for Sleep Tracking on Apple Watch as well as a Health Checklist for managing safe practices functions (Emergency SOS, Medical ID, Fall Detection, and ECG).

Thanks to Apple's acquisition of Dark Sky, severe weather alerts, a precipitation forecast for the next hour, and minute-by-minute precipitation readings are now included in the weather app.

Bicyclists and commuters can use the *Apple Maps app* to locate bicycle routes that take elevation, traffic, and the presence of stairs into account. In the event that you have a personal electric vehicle, you can choose a route that is tailored to your vehicle and charger type.

A well-curated guide to the best new restaurants and activities in a city provides a list of places to go and things to do. A wide variety of manuals are made by well-known companies, *including the Washington Article, All Trails, Complex, and more.*

Users can use their iPhone or Apple Watch to unlock or start their vehicle with digital keys, and after a year, the U1 chip in the car keys will allow customers to open their vehicles without removing their iPhone from their wallet or luggage. Text messages may be used to share vehicle keys, and iCloud could be used to disable them if an **iPhone** is misplaced.

Customers can select their own wallpapers in CarPlay, which also aids in the development of new apps for parking, recharging electric vehicles, and placing takeout orders.

Adaptive Illumination allows Home Kit lights to modify their color temperature during the day using the Home app's automation ideas and Handle Center quick access buttons.

Home Kit Security Video cameras help Activity Areas for the first time by using on-device Face Recognition to identify who is at the door (based on recorded images of individuals).

Text and voice translations are now available in 11 languages thanks to a brand new Translate software from Apple. Translating on a mobile device can be made easier using an On-Device Setting that allows just translations to be downloaded. The Discussion Mode speaks translations aloud, allowing users to converse with someone who has a different lexicon and have the software accurately translate their words as they are said.

Arabic, Chinese, British, French, German, Italian, Japanese, Korean, Portuguese, Russian, and Spanish are only some of the supported languages.

There are new options for restricting the use of location data, such as allowing apps to access only the approximate location of a user's device, in the *latest privacy protections.* New symbols appear on the home screen when an app intends to use a camera or microphone and needs user permission to monitor them across online pages.

What is the Importance of iOS 15's New Protection Features for Your iPhone?

Apple's strategy for distinguishing the iPhone from the plethora of Android devices on the market is based on privacy. Several new features are included in *iOS 15*, which further solidifies Apple's position as the undisputed leader in this field.

There are various privacy-focused features in *iOS 15*, which is currently available in a public beta, that should help you have a much more secure smartphone experience.

These can be utilized by the normal person, and they aren't just for people who have to be concerned about their privacy in certain situations.

Are these new features going to change how you use your phone? Time will tell, but we've looked into some of the key features coming to your iPhone—and can likely be prepared for the iPhone 12—that might ensure it is safer to use than ever before.

Transparency is a big deal

Apps on your iPhone track you as you use them, just like every other technological system does. You need to know what these applications are tracking, and iOS 15 was designed to make it easier than ever to see what information they are collecting.

Apple is making substantial changes to the App Store to ensure that all apps have an inventory that lets you know exactly what this app monitors. This is especially true if the app claims to know your location and may read your phone number before you've even installed it on your phone.

Because of this, Apple is requiring all app developers to disclose this information in their app listings. A program will be denied entry to the App Store if it lacks these specifics.

Independent cybersecurity analyst Graham Cluley told TechRadar, "Anything that enhances transparency with respect to what apps normally do must be positive information for customer privacy, thus I appreciate this work." Graham said.

"If the usual consumer concentrates on permission notifications such as other issues completely." Everybody knows that many of us merely click through any permission screens without reading through *(or at the very least knowing)* the consequences of what we have just decided to.

The decision to utilize or not utilize this information will be up to the user, but even if you don't, other features in iOS 15 are already boosting other apps' components.

Whenever you paste from another device *(e.g., copy on a Mac PC and paste on an iPhone)*, the **iOS 15** beta displays a verification banner.

When TikTok learned about the privacy problem, they pledged to fix it, which should make the app safer for everyone, not just individuals who are concerned about their data being exposed.

Are there going to be fewer apps as a result of this?

There's a good chance that fewer third-party apps will be

submitted to the store as a result of TikTok, because not every corporation will necessarily receive the same safeguards.

Having said that, it's safe to assume that Apple will be content to lose the small but vocal group of designers who don't adhere to these standards of conduct. Cluley believes that most programmers will follow these recommendations as well.

There is no doubt, he added, "I'd guess any software that sought to sneak through excessive levels of monitoring or data selection without being upfront and honest about any of it with Apple and users would find themselves potentially facing analysis from the App Store."

- **Location improvements**

Using an app to describe your *"approximate position"* rather than your precise location is another security-focused feature in iOS 15. At this point, it's difficult to tell exactly what this means, such as how far away or how long ago your location was reported.

Most apps that require your location don't need your particular area, "**Cluley said**." *"For people's privacy and personal security, this can be a very logical stage."*

You may want to share your precise location with a slew of apps. Is it really necessary for social media apps to know more about you than just the area you're in, such as in the case of Apple Maps or City Mapper? Within iOS 15, you'll be able to make that decision.

What does this all mean?

Cluley points out that the new features in iOS 15 will only be as amazing as they appear if people start using them. There's no guarantee that the typical person will use these functions, but for those who want to be more aware of their surroundings, this can be helpful.

The status bar of your phone will light up with an indicator if an app is preventing you from using your microphone or camera, which is another new function. However, you won't know whether or not this is a problem unless you understand what that means.

Cluley believes that many privacy-conscious users will already choose Apple's operating system when asked if iOS is the preferred option.

"Android phones use an operating system that is operated

by a marketing corporation that makes money off of the information they can gather about their clients," he stated. Apple has set itself apart from the competition with its strategy.

Of course, selecting an operating system is only one part of the puzzle. In most cases, third-party apps and solutions pose privacy risks that are platform-independent, but ora can help raise awareness.

iOS 15's subtle changes ensure that it is just a little bit better and a little bit safer to use than previous versions, but they don't change everything on your iPhone.

Complete Five Actions On Start or Setup Wireless Carplay on iPhone

Setup your iPhone for wireless carplay by performing the following five steps:

1. The first step is to open the *Settings app* on your iPhone's home screen.
2. Navigate to *the General section.*
3. The third step is to press the *CarPlay button.*
4. You'll see a pop-up message if *Bluetooth* is

disabled on your iPhone at this time.

Start **Bluetooth** by pressing the Start button. If you're having trouble with Bluetooth, try **Repair BT** from the soft restart on your phone.

5. *In order to begin the **CarPlay setup** process, hold down the Tone voice control button on your steering wheel for a few seconds.*

Now that you've set up *CarPlay* wirelessly on your iPhone, While you're driving, use all of your iPhone's features.

- **Troubleshooting**

➤ As a first step, *go to Settings > Siri & Research to appear between your Wallpaper and Face ID/Tap ID/Security password tabs > Turn on Siri.* Toggle the switch for which it is possible to vote **"On/Green"** Once you hear the command *"Hey Siri,"* press the side button for Siri to invite her to join you.

➤ Verify that your car stereo supports *AirPlay*.

➤ Use the smartphone icon or the *CarPlay* icon to find out if your **CarPlay** is compatible with wireless and/or *USB* connections.

➤ Make sure the car is in motion at all times.

➤ **Hardware malfunctions**: Reboot the *iPhone*, swap out the *USB connector*, and install the most recent version of *iOS*. A stereo system's firmware is also regularly updated.

Chapter 13
iPhone 12 Cheat Sheet

When it comes to the new iPhone 12, what are the must-have features?

5G Network:

Both sub-6GHz and wave systems are supported by the 5G technologies in each of the four new iPhone 12 models. It's even more widespread, can travel further distances, and is more resistant to interference than the sub-6GHz band. However, it is becoming even more congested, which limits its true speed. It's a lot faster, but it's also much more prone to interference, so it's most effective at a much closer range to cell towers. AT&T, T-Mobile, and Verizon Cellular all offer both types of systems in the United States, but Verizon's Wave is the one on the rise.

Carriers have begun rolling out their 5G networks, although only big cities in the United States have access. If you don't live in the appropriate place, you won't be able to get the best out of a 5G smartphone.

By including 5G, the iPhone 12 series will be more

resilient in the long term, since 5G coverage is expected to grow in the coming years. Budget-conscious smartphone owners are spending more time on their devices than ever before. Obviously, Samsung and other Google Android phone makers have already started incorporating 5G into their devices, so Apple needs the capability only if it wants to remain competitive.

An excellent Data Mode function was added by Apple since 5G could drain the battery. When 5G isn't needed, the phone immediately drops back to 4G.

Processor:

Processing power is provided by Apple's new A14 Bionic technology, which is also used in the *2020 iPad Air 4*, which was unveiled in September of that year. In contrast, the A14, Apple's new mobile processor, was designed to increase performance and save battery life. Four high-efficiency cores and two high-performance cores are included in the new six-core CPU chip.

This year's A13 Bionic chip has 8.5 billion transistors, but the upcoming 5nm chip has 11.8 billion transistors. As a result of the A14's increased transistor count, it is not just faster but also more efficient. The A14's neural motor can complete 11 trillion cycles per second, increasing the rate at which computations and device learning take place.

Display:

Apple's Super Retina XDR screen and *High Definition Range* are now standard on all new iPhones (**HDR**). They are some of the most advanced screens ever used in an iPhone, with a 2,000,000:1 contrast ratio and 1,200 nits of optimal brightness. When combined, these features produce better images, use less power, have more accurate colors, and have a far higher contrast ratio than typical LCD screens.

Ceramic Shield Screen:

The iPhone 12's front-facing cup display is made of ceramic, making it more durable and resistant to falls thanks to this function. The nano-ceramic crystals formed in the cup matrix are the result of a brand-new high-temperature crystallization method. Apple's tests show that the Ceramic Shield increases the iPhone 12's drop resistance by four times compared to the iPhone 11.

Rear Digital Cameras:

Both the iPhone mini and the iPhone 12 include dual 12-megapixel rear cameras, one with a wide-angle lens and the other with an extra wide lens. There are three 12-megapixel cameras on the rear of the iPhone 12 Pro and iPhone 12 Pro Max. The fourth zoom lens on all Pro models is used for **LiDAR** mapping, which improves the precision with which objects' range and level can be measured. This is very useful for augmented fact apps. It has an f/1.6 aperture, the fastest ever on an iPhone, which Apple claims gives 27% more brightness in low-light situations.

Both 12-megapixel telephoto cameras with a 52mm focal length, 4x or 5x optical focus, and optical image stabilization are available in both iPhone 12 Pro models.

There are no differences in frame rates between the rear cameras on the iPhone 12 and the iPhone 12, but the ones on the iPhone 12 Pro can capture HDR movies with Dolby Vision at 60 frames per second.

Front camera:

The iPhone 12's front-facing 12-megapixel camera has an

f/2.2 aperture and can capture HDR video at 30 frames per second with Dolby Eyesight.

Quick Take:

The Quick Take function was introduced in *iOS 13* on the iPhone 11 and allows you to record a short video while you're taking a conventional image. After that, you'll find a small clip of the video in your Photos folder.

Night Mode:

As soon as it's dark enough, this digital camera mode is activated and works in tandem with the latest sensor technology to produce stunning low-light images. In order to work better in low-light conditions, Night Mode requires shorter and longer structures and integrates them instantaneously. The Night Mode feature is available on both the back and front-facing cameras of all iPhone 12 models. Night mode Time-Lapse allows for longer exposure lengths, better lighting, and smoother direct exposure for time-lapse photography while your phone is on a tripod.

Smart HDR 3:

Using the Wise HDR feature, you can take superior HDR images and portrait setting shots with new digital camera sensors.

ApplePro Raw:

It allows you to take images in a green format but with all the latest advanced camera equipment and technology to enhance the image. This feature is included in both the *iPhone 12 Pro and the iPhone 12 Pro Max*. You can edit ApplePro Green photographs on the iPhone's built-in ***Pictures app***, as well as in third-party apps like *Instagram and Snapchat*.

Design:

As a result of a major redesign, the latest iPhone models have a flat edge design, which is similar to what was seen on the iPhone 5 and the iPad Pro before it.

MagSafe:

New to the iPhone 12 lineup is MagSafe, a wireless charging system that improves speed and precision. There's still more, however. Add-ons can be attached to the phone using ***MagSafe's magnetic foundation***. Apple, for example, has promoted a MagSafe-compatible slimline case that attaches to the phone's back.

Battery:

Because of the higher drain caused by 5G, the batteries in the *iPhone 12 models* will surely be larger than those in the iPhone 11 models by 10%. When compared to the iPhone 12 Pro and Pro Max, the iPhone 12 mini's battery life is much more constrained due to its smaller size.

Colours:

The iPhone 12 mini and iPhone 12 are available in a variety of colors, including the standard white and black, as well as blue, green, and reddish. Blue, precious metal, graphite, and metallic colors are all available for the *iPhone 12 Pro and iPhone 12 Pro Max.*

More Features on an iPhone 12

These features are also present in the iPhone 12 series:
- Wi-Fi 6 (802.11ax) offers download speeds that are 38% quicker.
- Improved water and dirt resistance: The iPhone 12

Pro has an IP68 rating for up to 4 meters of drinking water for 30 minutes.

- Fast charging to 50% battery capacity in 30 minutes.
- Using two different models of AirPods on the same iPhone to post audio.
- Spatial audio that simulates surround sound
- Dolby which simulates 3D motion in a 3D space
- Dual-sim phone based on esim.

Apple is offering the iPhone 12 without earbuds or charging plugs in the package in order to lower the cost of the new devices and reduce the environmental impact of accessories. The only accessory included in the box may be a standard USB-C charging cable. Apple may also assume that many customers already have earbuds and USB chargers on hand.

What are the Difference Between the Four iPhone 12 Versions?

iPhone 12 mini:

A miniature version of the iPhone 12 Apple's latest

iPhone, the iPhone 12, has a 5.4-inch screen with storage options of 64GB, 128GB, or 256GB. There are 2,340 x 1,080 pixels and 476 pixels per inch (PPI) on the screen. A ceramic shield in the shape of a cup and an aluminum shield are included in the mobile phone. The iPhone 12 mini has a dual-camera system. It has a 12 MP wide and a 12 MP ultra-wide zoom lens on the back. The phone starts at $699 and comes in black, white, red, blazing blue, and green.

iPhone 12:

The iPhone 12 is the next generation of the iPhone, and it has the same specifications as the iPhone 12 small. When it comes to pricing, though, the iPhone 12 comes in at $799 for the base model, which sports an AMOLED screen with a 2,532x1,170px resolution and 460ppi density.

iPhone 12 Pro:

The iPhone 12 Pro has the same 6.1-inch screen size and resolution (2,532 x 1,170 pixels) as the iPhone 12, but it includes a number of more advanced functions. The front side of the gadget is made of Ceramic Shield, and the back is made of a textured matte cup and stainless steel. A choice of 128GB, 256GB, or 512GB of storage is available. A 12MP telephoto zoom lens with an f/2.0 aperture and a LiDAR scanning device are provided by the three-camera system in the back. Apple ProRAW is also supported by the iPhone 12 Pro, allowing users to take and edit photos in the green format. The starting price for this model is $999.

iPhone 12 Pro max:

The iPhone 12 Pro max is the most powerful version ever. However, it's a 6.7-inch screen with a resolution of 2,778x1,284 pixels at 458 ppi that makes it the iPhone 12 Pro Max's crowning achievement. The optical and electronic zoom rates may also be slightly higher than those for the iPhone 12 Pro, while the price starts at $1,099.

What are the Main Competitors of the iPhone 12?

Your competition is wide open because the iPhone 12 comes in a variety of sizes and features at a variety of prices.

iPhone 12 mini competition:

the iPhone 12 mini, similar to the Samsung Galaxy S10e and the Google Pixel 4, which are 5.4-inch iPhone 12 minis. In addition to a 5.8-inch AMOLED screen and dual rear cameras (broad and ultra-wide), the Galaxy S10e starts at $599. It starts at $799 and features a 5.7-inch OLED display, dual rear cameras (one wide-angle, one telephoto), facial recognition, but no fingerprint sensor. There is no 5G connection on any phone.

iPhone 12 and 12 Pro competition:

There are a number of Samsung Galaxy cell phones that compete with the iPhone 12 and iPhone 12 Pro in terms of performance. Aside from that, only the S20 and FE are commonly 5G variants. Starting at $749, the Galaxy S10 has a 6.1-inch screen, while the S10 Plus has a 6.4-inch screen, making it the most expensive version of the phone. The S20 FE, which starts at $699, has a 6.5-inch screen, while the S20 starts at $999 and has a 6.2-inch screen.

Google's Pixel 5 and the impending Pixel 4A 5G, both of which have 5G capabilities, are worthy of consideration. At $699, the Google Pixel 5 is the most expensive phone in the Pixel lineup, with a 6-inch screen, while the Pixel 4A 5G starts at $499 with a 6.2-inch screen.

iPhone 12 Pro Max competition:

Samsung's Galaxy S20 Plus, Galaxy S20 Ultra, Galaxy Notice 20, and Galaxy Note 20 Ultra are the most common competitors to the iPhone 12 Pro Max. The Galaxy S20 Plus has a 6.7-inch screen and typically starts at $1,199. The S20 Ultra has a 6.9-inch screen and costs $1,399 when new. In comparison, the Note 20 Ultra costs $1,299 and has a 6.9-inch screen, while the Note 20 costs $999 and has a 6.7-inch screen.

Chapter 14
New iPhone Quick Fix

It's fun to get a deal on a pre-owned iPhone. When all is said and done, you will be able to buy an iPhone and extend your budget by purchasing a used one, especially for those who are less financially secure.

Some people run into this problem when trying to turn on a new gadget: For security reasons, the new iPhone usually won't work without an Apple ID.

Don't worry, because if you follow these steps, you'll be able to solve this problem:

- A security mechanism introduced by Apple in response to the growing number of iPhone thefts is called *Activation Lock*, and it is part of the company's *Find My iPhone program*. It used to be possible to clean, resell, and get away with stealing an iPhone without it being protected by a lock mechanism. The scenario was drastically altered by the activation lock.
- The **Apple ID** used would be kept on Apple's activation servers along with nearly every other piece of information about the phone after the

initial owner setup detects my iPhone on the mobile phone. If a unique **Apple ID** can be used to unlock the phone, the activation servers will be most effective. In the event that you lose your **Apple ID**, you will be unable to activate or use the phone. It helps keep your iPhone safe because no one wants to take a phone that they can't even use. On the other hand, if you just got the phone, it's probably not a problem.

- In addition to being irritating, the activation lock can be easily solved. It's a possibility, but it's more likely that the previous owner neglected to turn off or erase the tool before selling it on the market.
- You should get in touch with the previous owner of the phone to see if they'll take the appropriate measures.

iCloud Activation Lock Removal

When the merchant or seller is unable to use the phone due to, for example, distance, things might get a little confusing and complicated. In order to remove the activation lock, follow the steps outlined below in order to access iCloud and remove the lock on the phone:

- Use any device to access iCloud.com, including a mobile phone or laptop.
- Access the phone's settings using the *Apple ID* that was used to activate it.
- Go to the *Find My iPhone* page.
- Select the *All Devices* option.
- Examine the iPhone you've sold or wish to sell.
- Click on "***Remove from Accounts.***"

Once you've done that, you can unplug the iPhone and turn it back on. Next, you can proceed with the regular activation process.

How to Remove Activation Lock on iPhone

- You should unlock or remove the activation lock from the iPhone you just purchased by entering the previous owner's **Apple ID**. Getting in touch with the business owner and explaining the situation is the first step in this process.
- It is best if you can return the phone to the owner in person and have him or her enter the required

unlock code, which is his or her **Apple ID**. When the buyer receives the iPhone, the only thing they have to do is input their **Apple ID** on the activation lock screen. After doing so, restart the phone and proceed with the usual activation process.

Using the Find My iPhone app, here's how to format an iPhone.

In fact, this method is nearly identical to the method described above, utilizing iCloud, by just accessing the Find My iPhone app on another iPhone. Make sure the phone is connected to Wi-Fi or mobile data, and then tell the owner to follow the instructions below:

- Go to the *find my iPhone* app.
- Log in with the *Apple ID* applied to the phone sold to you.
- Choose the Phone.
- Click *Actions*.
- Click *Erase iPhone*.
- Click *Erase iPhone* (It is the same button, however, on a new display).
- Enter *Apple ID*.

- Click *Erase*.
- Click *Remove from Accounts*.

Start the setup procedure by restarting your iPhone.

How to Wipe an iPhone Using iCloud

If you are unable to approach the vendor/merchant for whatever reason, but still need your mobile phone to be cleaned completely for convenience, the seller may use *iCloud* to remove it. Ensure that the phone you want to buy is connected by WiFi or mobile data, and then tell the vendor to proceed as follows:

- Visit https://iCloud.com/#find
- Sign in with the *Apple ID* he/she created on the phone you have or that he/she purchased for you.
- From the drop-down menu, select *All Devices*.
- Choose the phone that you will buy or have access to.
- Choose "*Erase iPhone*."
- When the phone has been deleted, select *Remove from Accounts*.
- Restart the phone and you're good to go.

Chapter 15
Top Recommended iPhone Apps

Spark: Best Email App for iPhone

Those who focus on iOS apps will notice that email has taken on the role of the antagonist in the iOS universe. A better email platform and a credit card application may be obvious to app developers, but they may not be aware of this fact. If you're using Spark, you'll be able to send, snooze, and have an inbox that only alerts you to important emails, making email management a little less stressful.

Things you'd like concerning this application::

- A swipe-based interface allows for one-handed use.
- The software is easy to use and friendly to the community.

Things you might not like:

- For example, there are no automatic filtration mechanisms for sorting electronic mail and the software does not allow you to manage messages in bulk.

"To-do manager" for the iPhone: Things.

Things isn't the only good to-do manager program, and it's not the only to-do manager on this list, but it's a cautiously trustworthy tool that sits between control and hardiness. A well-balanced application form gives users just the right amount of control and toughness, but doesn't overwhelm them with dials or leave out important options.

Things you'd like concerning this application:

- This software has a simple UI that lowers stress when adding and completing tasks.
- Using the sheet extension, tasks can be added from iOS.

Things you might not like:

- If repetition and deadlines annoy you, you should avoid this program.
- It is not possible to automatically schedule tasks.

OmniCentre is the bestiphoneGTD-compliant To-Do app.

Like **"Things,"** OmniCentre is a widely used task

manager with an attractive interface, although its priorities are different. Things is a strong and feature-rich alternative to OmniCentre, which aims to keep things simple.

Users are encouraged to fill in any tasks they have and their associated information and scheduling in the application form, utilizing the "Getting Things Done" (GTD) technique of task management. If the program doesn't have enough features to implement the entire GTD process, users will wind up spending a lot of time on lead closing organizing work.

Your favorite features of this application include: the best to-do list manager available; can be used with virtually any task management method. "

Things you might not like:

To get power and versatility, you'll have to sacrifice ease of use.

Agenda: Best iPhone App for Busy Notice Takers

In contrast to other note apps, Agenda is a "date-centered notice taking app," or "notes app." Changing times are a big part of the agenda, and records are organized by job and day. A to-do list is created from "things" rather than a collection, making Agenda an efficient journaling software and an effective to-do organizer and general iPhone note-taking application. Agenda may be the first iOS note-taking app to perform this combination efficiently, despite the fact that your day and notes seem obvious.

It's a "to-do manager" in addition to a note-taking app with some calendar functions, allowing you to see all of your information in one location and from one perspective. Freeform application forms can be really useful, which may not be typical in flagship applications. When using "Agenda" with pencil support, the app's magic comes to life, but for the time being, we'll have to settle for the iPad Pro's version of the feature.

Things you'd like concerning this application:

- Note-taking simple modifications can improve many workflows, which is something you'd like to see in this program.
- For the majority of users, the time-based structure works best; mental forms of information storage are preferred.

What you might not like:

- Slow app releases can limit your ability to

write down a note swiftly.

1Password: The best password management app for iPhone users.

As far as password managers go, *1Password's* auto-fill on iOS 15 is the best we've seen. However, access to Face ID makes the application form easier and better to work with, which is a rare combination of successes to achieve simultaneously.

Advantages:

- Username and password copying is a breeze thanks to this program.
- Auto-fill functionality finally makes account password management as simple as inputting your account password.
- Protected document space for storage means that 1Password can gather all your secure data in one location.

Disadvantages:

- There is no free version, and the paid version is based on membership.

Twitterific: Best Tweets App For iPhone

Even though Twitter isn't the best social networking platform, it's still one of the most popular websites, and the default app for the site is a disappointment.

Third-party Twitter clients have recently been savaged by Twitter. It appears that this move is intended to force users into using the native app; however, given the program's numerous flaws, Twitterific and the third-party applications that utilize it are still preferable.

Advantages:

- In terms of this app, it substantially improves the visual presentation of Twitter.
- It has wise and strong features that make Twitter easier to use.

Disadvantages:

- Some organizational choices are initially unintuitive.
- Twitter has deliberately kneecapped a vast number of third-party programs, and Twitterific is usually no defense to the consequences.

Overcast: Best iPhone App for Podcasts

Overcast may be the best podcast player you can get your hands on. "Smart Rate" and "Modulation of voice Boost" are two features of the app that are designed to make podcasts more enjoyable to listen to in a noisy environment, while "Smart Rate" intelligently adjusts a podcast's playing speed in order to reduce silences without accelerating speech.

Some things you'll appreciate about this app:

- A well-designed interface for organizing and listening to podcasts;
- useful features such as Smart Speed and Queue playlists;
- and a developer who is concerned about not exploiting the app's users for financial gain.

Disadvantages:

- When it comes to the iOS lock screen, it doesn't appear to work well.

Apollo: Best iPhone App for Reddit

If you're thinking about signing up for Reddit, you'll need to go beyond the app. The application form has improved, but third-party offerings are still miles ahead of it.

When it comes to Reddit users, Apollo has surpassed previous champions like "Narwhal." The app's development is on-going, as seen by the multiple updates posted by the app's developer on Reddit.

As long as the iPhone application switching behavior is followed, the swipe-based navigation will continue to focus on any iPhone. For OLED screens, a true black setting can be a delicacy.

Useful features:

- An easy-to-navigate UI makes it simple to find what you want.
- Effortlessly handles a large selection of media.
- There is no advertising in any version of the application.

Disadvantages:

- Sometimes there are bothersome and lasting bugs that are difficult to get rid of.

Focos: Best iPhone App for Editing and enhancing Portrait Setting Photos

Face Mode on the iPhone is a one-and-done process; you

take a picture, and the blur is applied automatically. This is a fact: there is no built-in technique in iOS for tweaking and improving the picture setting effect. A tool for adjusting both shadows and the blurred nose and mouth masks is created when Focos fills the space. It simulates the effect you'd get by adjusting the physical aperture of a zoom lens. Incredibly, you can even re-create the blurred background on other objects, or even yourself, by changing the effect within the image's depth, nose, and mouth mask in a matter of seconds.

Advantages:

- Portrait Mode's depth-of-field effect can be tweaked in this app's most reliable way.
- As a distinctive feature, depth maps can substantially aid in visualizing blur.

Disadvantages:

- Over-processed photos are easy to make, and only about half of the blurry stove looks natural in the center.

Halide: Best iPhone App for taking Natural Photos

Halide is unique in that it embeds critical information in the iPhone's "ear." Can it be treasured because it has a live histogram for image evaluation? It's not quite flawless, but Halide is still a near-perfect photo-taking program.

Setting and configuration are great, the RAW catch is pixel-perfect, and the application form is easy to use and navigate. Halide may be the greatest camera software for iOS if you want to take images on your iPhone.

Among the features you'll appreciate:

- low processing power for iPhone photos.
- The widest range of tools for iOS image editing and enhancement apps.

Disadvantages:

- First-time users may feel overwhelmed by the lack of control they have over the system.

Euclidean Lands: The Top-rated AR Puzzle Game for iPhone

The killer app for augmented reality has yet to be discovered. AR games, on the other hand, require the use of a lot of iPhone features in order to get the most out of them.

Euclidean Lands is a short but enjoyable puzzler that

takes full advantage of the AR's capabilities. By manipulating the play space and creating new paths through the puzzle patterns, players in Monument Valley can lead their avatar to the end of the maze. The game starts out simply, but you may find yourself a little stumped near the end.

Advantages:
- Puzzles levels that are challenging and visually appealing make use of AR's unique features.

Disadvantages:
- The primary game repair shop feels overly familiar.
- Very short.

Giphy World: Best AR Messaging App for iPhone

Snapchat's AR messaging mechanism has been usurped by numerous other applications. While Snapchat may be in a weaker state as a result of its own actions, it is far from being destroyed. If the number of users decreases, Giphy World is an excellent replacement.

You'll appreciate the following features of this program:
- Easy creation of amusing images using

available assets;
- Content is not restricted to use within the Giphy app.

Disadvantages:
- Object location and processing speed are inferior to Snapchat's.

Jig Space: Best Using AR for Education on iPhone

Learning through holograms is a common theme in science fiction films, but thanks to Jig Space and augmented reality, it's now a reality in our everyday lives, as well. By using the app, you can learn other things about the system by using the app, including how each lock type works, modifying every part of the system, and looking at it from different angles. Jig Space necessitates the use of AR's three sizes and the low-poly models to ensure that the quality of the visualizations is never compromised.

Advantages:
- it takes advantage of AR's benefits for a good cause and provides a wide variety of

"jigs" for free.

What you may or may not like:

- Occasionally, the captions that accompany the images are sadly shallow.

Nighttime Sky: The Best Late-Night Companion App

More fun is had when you aren't creating the constellations as you walk. On iOS, Evening Sky was the first augmented reality-style app. Although others on the computer have tried to duplicate its success, it has remained dominant.

Things you'd like concerning this application:

- It enhances the natural world with technology. It improves the stargazing experience for both youngsters and adults.

What you might not like:

- Because of the big image units, the camera's actions seem stiff and unnatural.

On iOS, Inkhunter is the best AR gimmick.

The experience of researching new tattoos on your own has a special allure. uses augmented truth to produce temporary digital symbols that you can place on your body and take a snapshot of. When you want to project on your skin, you need to use the built-in Adobe Flash, pull your designs, or import property from another source.

Things you'd like concerning this application:

- It's a fun and novel application idea that's beneficial.

What you might not like:

- Having to deal with AR's current surface matching constraints.

Chapter 16

How to Format iPhone 12

Here are some steps on how to reset the iPhone 12's operating system to factory settings. Please keep in mind that these methods may not work if your iPhone is experiencing additional issues.

In order to perform a hard reset on your iPhone 12 via the Software Program Menu, follow these instructions:

- The battery should be charged properly.
- Turn on *iPhone 12*.
- If you haven't already, please utilize iTunes to quickly backup and *restore* all of your crucial data.
- Select *General* from the Settings menu.
- Please go to the bottom of the *General* menu page and select *Reset*.
- Apple's *iPhone 12* can be *hard reset or reformatted* by selecting the option to *Erase All Content and Settings*.
- Please wait a few minutes before using the device in its factory default state.

A PC-Based Hard Reset of an iPhone 12 (You can also

use these options for iTunes backup and restore):

- *Download* and *install* the iTune software on your PC (MacOS or Microsoft Home Windows).
- Connect your iPhone via USB cable to your own computer after iTunes has been set up.
- On *iTunes*, you must *select* the *iPhone* as the *device*. If our iTunes appears on the sidebar, we can also use the gadget option to access them.
- *Restore* should always be selected.
- You'll be able to use the software to keep your computer up-to-date with the latest Windows updates.
- In the event that you come across a software license agreement, select *Agree*.
- You'll have to wait until iTunes displays "*Welcome to Your New iPhone*" before you can use the new iPhone 12.
- To use the new iPhone 12, choose *Setup a brand-new iPhone*.
- If you wish to restore all of your iPhone 12's data from a previous backup, select *Restore*.

How to Restore and Reinstall, Clear iOS/Firmware on iPhone 12

Your iPhone 12 should run well on iOS 12 if you don't encounter any problems. In spite of this, an application accident can cause iOS damage. You'll need to reformat or reinstall the stock iOS on your device. To begin with, you need to know that the iOS document source is already safely stored in the iPhone 12 ROM or memory, so you don't need to copy it from another device unless you reformat. Instead, you should utilize iTunes to back up and restore all of your data. The clean manufacturer's default iOS firmware or operating system will surely be installed once you stick to the instructions to hard reset or reformat above.

How to Unlock or Fix or Bypass Forgoten Security Password

Screen lock protection on the iPhone provides a high level of security for our personal information. When your iPhone 12 is dropped, taken, or stolen, you can rest assured that your vital or confidential information is safe. When you lose track of your security pin or password, you'll run into trouble.

How do you bypass or unlock the iPhone 12's security or password protection?

In addition to using iTunes to reinstall or restore, you'll need to do a hard reset and delete all of your vital data. The methods for performing a hard reset with iTunes previously described in this book can be used to recover any previously saved data.

How to Make a Backup of Your iPhone 12 Data and Restore It

In order to back up and restore all of your vital data and installed programs, as well as the device menu, you'll need to download and install iTunes on your MacOS device (Microsoft Windows Operating System). The procedure to back up and restore your phone should be undertaken after you have performed a hard reset using one of the methods described above.

How to Update iOS at iPhone 12

You can download the most recent version of iOS via

OTA or even via iTunes if you choose.

How do you enter Accessibility mode on iPhone 12 and 12 Pro!

In contrast to other gestures, you'll need to plan ahead for this one.

- Launch *Settings* from the *home screen*.
- Tap on *general*.
- Tap on *accessibility*.
- Turn accessibility to **On**.

Afterwards;

- Tap your finger on the bottom of the *iPhone 12's screen* in the gesture area.
- Swipe down.
- The *Control Center* may be accessed by swiping down from accessibility's upper right corner.

Chapter 17

iPhone Tips & Tricks

How to Enable USB Restricted Setting on iPhone

The latest version of iOS from Apple includes a powerful new security feature known as *USB Restricted Setting* for the iPhone. Devices that can be plugged into the USB port on an iPhone and use the device's passcode have recently been developed by several companies.

Apple has implemented a USB Restricted Setting in order to prevent this. The USB Restricted Setting prevents data transfer between an iPhone and a USB device if the iPhone has been locked for more than one hour. This renders iPhone unlocking boxes useless because it may take hours or even days to unlock a locked iPhone.

Apple's mobile operating system, iOS, has *USB Restricted Mode* turned on by default. In order to turn off Face ID and Passcode, open the *Settings app* and tap on *Face ID & Passcode*. Swipe down after you've entered your passcode until you see the option to *"Allow Access When Locked."*

"USB Accessories" is the final toggle in this section. A *USB Restricted Setting* is allowed, and devices can't download or upload data from/to your iPhone if the iPhone isn't unlocked to receive more than an hour of data.

Use Your iPhone to Control Your Apple TV.

If you have an Apple TV, you can use the Control Focus on your iPhone to control it. The only requirement is that both your iPhone and Apple TV be on the same mobile network. Look for the Apple TV button in the Control Center after you've accessed it. To begin managing your Apple TV, simply touch the button.

Make use of the Two Pane Scenery View option.

The information in this advice is specific to the new

iPhone 11 Pro Max, but it's still worth knowing. Email and Records are two of the built-in apps that convert to a two-pane layout when the iPhone XS is horizontally oriented. You can see a list of all of your notes in the Records app while actively reading or editing a single note in this main iPad setting.

Turning Off iPhone Alarms by with your face

Face ID is one of the iPhone's most intriguing new features. You'll be able to unlock your phone by looking at it. Apart from that, Face ID has a slew of additional intriguing capabilities. It's possible to silence your iPhone's security alarm by simply picking up your phone and looking at it; this shows your iPhone that you're aware of the alarm, and it'll quiet it.

Disable Face ID in a flash.

The authorities may be able to use face recognition technologies on your smartphone to legally demand that you reveal it at that moment, depending on where you live. For reasons that remain a mystery, facial biometrics are not always safeguarded in the same way that fingerprints and passcodes are. Because of this, Apple has developed a characteristic that allows you to disable *Face ID* without having to go into your settings. By pressing the side button five times, you can turn off *Face ID* and unlock your phone with your passcode instead.

How to Slow the two times click necessary for Apple Pay

Using *Face ID* and double-clicking on the medial side button on the iPhone, you can verify your ***Apple Pay*** commitments. By default, you'll have to hit the medial side button twice as fast as normal to slow down the action.

Select Settings > General > Availability from the main menu to get started. The side button may now be found by swiping down the page. You have three options on the Privately Button screen: *default, gradual, and slowest.* Pick a speed that works best for you and stick with it.

Chapter 18
iPhone 12 Problems & Solution

Both Apple's iPhone 12 and iPhone 12 Pro have been plagued by a number of faults since they were released earlier this month.

Now that the iPhone 12 and iPhone 12 Pro have already been available for a few weeks, we're collecting feedback from those who got their hands on one early.

The majority of the feedback is still positive, but we've also learned that both phones have a number of performance and bug issues.

Fortunately, there's a strong chance you can fix your broken phone before new software is released or before you have to call Apple's customer service department.

To help you out, we'll go over some of the most prevalent iPhone 12 and iPhone 12 Pro issues. The list includes solutions for Wi-Fi, Bluetooth, and other issues, as well as a variety of other updates.

How to Improve the Life of the Battery on an iPhone 12

The battery life of the iPhone 12 and iPhone 12 Pro has been praised by several consumers. Others have noticed that the battery drains more quickly than predicted.

Prior to contacting Apple support, follow these actions if the battery life on your iPhone 12 or iPhone 12 Pro starts to degrade faster than expected.

It is imperative that we get customer feedback on the iPhone 12 and iPhone 12 Pro now that they are available to the public.

Most of the ideas are still excellent. However, we're also hearing about a number of issues that 6.1-inch owners are experiencing.

However, many users report that their electric batteries are depleting faster than they should, even though we haven't experienced widespread battery difficulties (at least not yet).

Complaints about 5G draining batteries faster than LTE aren't shocking because battery life issues are common (especially after Apple publishes new versions of its iOS software).

There are a few things you can do if you notice that your battery is draining rapidly.

Here, we'll walk you through a few tweaks that may help you get your iPhone 12's battery life back on track.

These are methods that have worked for all of us throughout time and can help you cure your battery troubles in a matter of minutes and avoid a conversation with Apple customer care.

Restart Your Phone

We usually recommend restarting your cell phone before doing anything else if your battery is dropping faster than you expect.

Wait one minute before turning on your iPhone 12 or iPhone 12 Pro. Take these further steps if it's still draining quickly.

Up-date Your iPhone

Apple releases iPhone software updates on a regular basis. Point updates (x.x.x) tend to focus on fixing issues, whereas milestone updates (x.x) tend to include a wide range of new features and fixes.

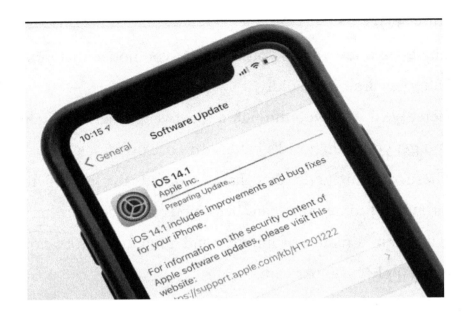

Even if Apple doesn't explicitly mention battery life fixes in an iOS update's release notes, new software has the potential to significantly improve battery life.

Turn off 5 G wireless technology.

Both the iPhone 12 and the iPhone 12 Pro support 5G. While speedy, it may deplete your battery far more quickly than LTE. If you don't need 5G or won't be using it much, you may turn it off in the iPhone 12 settings.

There are a few 5G setups in iOS 15 that you may want to consider purchasing as a memory lender. Head to *Configuration, then Cellular, then Cellular Data Options,* and finally Tone of Voice & Data.

In this case, there are three choices available:

- *5G On*
- *5G Auto*
- *LTE*

Even though it drains your battery, your iPhone will use **5G** anytime it is available, even if it does so without your permission.

The **5G Auto** option only activates **5G** when it won't severely drain your phone's battery life. It's possible that **5G Auto** will be the default and the selection that the majority of users choose.

Your Apps Need a Checkup.

If you're experiencing an unusual drain on the life of your phone's battery, you should take a closer look at the apps you're using.

This is one of the best ways to see how well an app is doing on the iPhone 12/iPhone 12 Pro. Detailed instructions are provided here.

- Open *the Settings app*.
- Scroll down to the *Battery Utilization tool* and select it.

In order to find out which apps are draining your iPhone

12's battery the most, you can utilize this battery utilization tool.

If you use an app frequently, it will deplete your battery faster than infrequently used apps. If a program that you rarely use is consuming a lot of power, you should look into it further.

It's worth testing to see whether uninstalling the program that's draining your battery life the most is effective at fixing the problem.

You should always get the most recent update from your app's creator if it is essential to your daily routine. However, if none of that helps and you still use the app to carry out your daily tasks, you may have to revert to an older version of iOS 15.

Reset All Settings

In order to avoid forgetting known Wi-Fi networks, make sure you have your passwords on paper or stored elsewhere before you do this.

- Go to the *Settings*.
- Click on "*General*."
- Scroll and tap *Reset*.
- If your pass code is enabled, select *Reset All*

Settings and enter it.

Reconnecting your iPhone to your Wi-Fi and Bluetooth systems and devices is required after the procedure is complete.

Switch to a low-power mode

iOS 15's *Low Power Mode* can help you conserve battery life by disabling solutions *(Hey Siri, automatic downloads, and email fetch)* that could deplete your battery.

When your iPhone approaches 20% power, the operating system may prompt you to carefully turn on the *Low Power Setting*, which you can toggle on or off at your discretion.

If you haven't already done so, Low Power Mode can be added to your Control Center. You may access the Control Center by swiping from the bottom of the screen or from the top right of the screen, depending on your iPhone model. This is how you do it:

- Open the *Settings* menu.
- Tap the *Control Center*.
- Click on the *Customize Settings* button.
- Press the green (+) plus symbol near the *Low*

Power Setting to access the further options.

You'll notice a battery icon when you open *Control Center* on your phone for the first time. By swiping it, the *Lower Power Mode* can be enabled or disabled.

It's even possible to activate the *Low Power Setting* in your *Settings* menu. This is exactly how you do it:

- Go to *Settings*.
- Tap *Battery*.
- Tap *Low Power Mode*.
- Turn it *On*.

Stop Background Rebrand-new apps

Rebrand-new apps should be stopped from running in the background, so they can show you the most recent data when you open them up. Although it's beneficial, it also drains battery power. It is possible to turn it off if you don't need it at all. This is exactly how you do it:

- Head to *Settings*.
- Tap *General*.
- Tap *Background App Rebrand-new*.
- It's best to turn it ***off***, as well as any unnecessary programs.

If you don't want to go through each app one by one, you

can even turn them off.

Downgrade

If you can't find a solution to the problem and don't want to wait for Apple's next iOS 15 upgrade, you can try downgrading your iPhone 12's software (if the choice can be acquired).

How to Repair iPhone 12 Missing 5G

After opening your new iPhone 12, you're undoubtedly wondering if your carrier has a 5G network that you can make use of. It is the first time that the iPhone 12 and 12 Pro have a 5G connection. That being stated, you should be working on an innovation that improves 5G connectivity.

It's not a glitch if you go to the Cellular section of the Settings app on your phone and there are no options for 5G. The most likely reason for this is that your service provider does not yet provide a 5G-enabled package. As a result, you will only see "LTE" and "4G" on your screen.

In order to take advantage of your carrier's 5G network,

you'll probably need to change or upgrade your approach. Make sure that 5G service is available in the places you usually visit before making a decision. These are the 5G provider maps you can use to do that:

- *AT&T 5G Map*
- *T-Mobile 5G Map*
- *Sprint 5G Map*
- *Verizon 5G Map*

One other note: You may also notice that your phone is connected to 5G even when you aren't using Wi-Fi on an *AT & T iPhone 12 or 12 Pro. AT&T's 5G* network isn't being used in this scenario.

The carrier's LTE-A support has been rebranded as 5G Development.

How to Fix Wi-Fi Issues

There are a few things you can try before contacting customer service if you notice sluggish Wi-Fi speeds or an increase in dropped calls.

First, check the Wi-Fi connection that is causing you problems on your iPhone 12.If you are using your home Wi-Fi system, unplug your iPhone.router for a few seconds before plugging it back in.

The router may not be the problem, but you may want to check to see whether other individuals with the same ISP are having similar problems in your area.

To access your iPhone 12's Settings app, if you can't access your ISP/router or if the problem is not related to your ISP/router, go to the Settings app.

If you have any difficulties once you arrive, be sure to check the Wi-Fi system. This is exactly how you do it:

- In Settings, tap *Wi-Fi*.
- Make a choice of connections by tapping on the *"I"* in the circle.
- Tap *Forget this network* at the top of the screen. *It's important to keep in mind that doing so may cause your iPhone to seek your Wi-Fi password.*

When all else fails, try resetting your iPhone's system settings:

- Head to your *Settings app*.
- Tap *General*.
- Tap *Reset*
- Tap on *Reset Network Settings*.

How to Fix Bluetooth Issues

You can take a number of measures if your iPhone 12 or iPhone 12 Pro is having trouble connecting to your Bluetooth devices.

Your Bluetooth connection may be the first thing you need to fix. This is exactly how you do it:

- Turn on Bluetooth by going to the Settings app and then tapping on Bluetooth.
- Click the connection utilizing the *"i"* within the circle
- Tap *"Forget this device"*.
- Try reconnecting to the *Bluetooth* device.

After you've tried that and it didn't work, try resetting your network settings:

- Go to your *Settings*.
- Tap *General*.
- Tap *Reset*.
- Tap *Reset System Settings*.

It will take a few seconds for this procedure to be completed. Make sure you have your Wi-Fi security passwords handy if you do this.

Even if all else fails, you can attempt to restore your

device's settings to their original factory settings. This is exactly how you do it:

- Go to the *Settings app.*
- Tap *General.*
- Tap *Reset.*
- Tap *Reset All Settings.*

If these workarounds don't work, you may need to contact Apple's customer service or the manufacturer of the Bluetooth device you're trying to connect to, depending on the nature of the problem.

iPhone 12 Charging Problems: How to Fix It

There have been reports of trouble with the iPhone 12's wifi charging, which we believe is to blame for many of these occurrences.

Restart your iPhone if you're having trouble with wireless charging. Press and launch to complete the task at hand. Before the cell phone shuts off, press and release the volume up and down buttons, then hold down the power button. If the feature is working properly, you can put the phone back in your pocket.

In order to charge your phone, you'll need to remove any

bank cards or security measures that are stored in the cover. As a last resort, you might unplug your gadget and see if it will still charge.

How to Fix Cellular Network Issues

If your iPhone suddenly displays a "*Simply No Service*" icon and you are also unable to connect to your cellular network, here are some measures you may take to fix this problem.

To begin with, check to see if there is currently a local outage. Do some research on social media to see what people are saying about your business and/or interact with your firm there. It's even possible to see if anyone else in the area is experiencing the same problem.

The only way to determine if a system outage is the source of the problem is to restart your iPhone and see if that resolves the issue.

Make sure to turn the **airplane setting on** for 30 seconds before turning it **off** if that doesn't work.

If that still doesn't work, you may want to try turning off cellular data altogether. What you need to do is as follows:

- Go to *Settings*.

- Tap *Cellular*.
- Turn *Cellular Information* to *Off*.
- Take a minute to turn it *off* and then turn it back on again.

How to Fix Sound Issues

The speakers on your iPhone 12 should be loud and clear. Here are some things to try before contacting Apple customer support if your audio begins to crackle or sound muffled.

Restart your iPhone first. It's also a good idea to make sure that your SIM card is in the correct position in the holder. The iPhone 12's SIM card slot is located on the left side of the device.

Even if you don't want to use Bluetooth, you can toggle it on or off.

Make sure there aren't any particles clogging the speaker grille or the Lightning port if the phone's sound continues to be distorted or missing.

Restart your phone if you notice a sudden decrease in contact quality. The recipient of the smartphone will also need to be checked to make sure it isn't obstructed by dust or a screen cover. If you're using one, you can even try to eliminate your scenario to see if that helps.

Try reactivating your phone's microphone if it suddenly stops working or starts randomly deleting.

It's possible to restore your phone from a backup if the broken mic persists. Apple should be contacted if the repair doesn't work and you suspect a hardware issue.

How to Repair iPhone 12 Activation Problems

If you're having trouble activating your iPhone 12 or iPhone 12 Pro, here are a few things you can try.

To begin with, check to see if Apple's systems are up and running normally. You can do so directly here on the company's Program Status page. If you see a green ring around iOS device activation, everything with Apple should be working as it should.

Make sure your SIM credit card is correctly placed in your iPhone if you see a natural symbol but still aren't able to activate it. Make sure you're using the correct SIM card as well.

There are a few additional steps you may take if you're getting an error message that says *"Invalid SIM" or "Zero*

SIM.":

- Check to see if you have a current wireless contract.
- Make sure you have the most recent version of iOS on your iPhone.
- *Restart* your phone.
- Make sure you've got the most recent version of your carrier's settings. Go to *the Configuration > Common > About page.* If an upgrade is available, you'll be given the option to **accept** it or **update**.

Contact **Apple** or your service provider if none of these solutions work for you.

How to Repair the Performance of an iPhone 12

Fixes for difficulties with slowness, freezing, and locking can be found here if your iPhone 12 or iPhone 12 Pro is experiencing any of these problems.

We're listening to a variety of complaints as we get closer to the **iOS 15.1** launch date, including a number of performance issues that could affect iPhone purchases.

iOS 15.1 has received mostly positive reviews, and we've been using a decent face across most of our products, but we've been hearing about bugs and performance issues.

Lag, lock-ups, and reboots are all on the list right now. Especially on older models of the iPhone, they're all too common. In terms of shape, newer iPhones aren't completely safe.

Unfortunately, it may be difficult to fix performance issues like these. You may be able to resolve certain issues yourself, but others may necessitate a fresh software update from Apple.

If you don't have the time or energy to get in touch with Apple support, we've got some fixes that may help alleviate your device's general performance issues.

Reboot your iPhone.

Try rebooting your iPhone if you notice UI slowness or other efficiency issues. As soon as the power goes out, turn it back on for one minute.

Up-date Your iPhone

RevisedApple's iPhones and iPads may, from time to time, release updates to the iOS 15 operating system.

Fixing issues is a common goal of point updates, but adding new features and functionality is the goal of milestone updates (x.x).

Update App

iOS 15 assistance updates are being released by app developers, and they may help to stabilize the overall performance of your device on iOS 15.

Be sure to read other people's reviews and critiques of iOS 15 and iOS 15.1 in the App Store before making the decision to upgrade. Even if the reviews are mostly positive, it's a good idea to upgrade to the most recent version.

Reset Your Settings

If you're still not getting the kind of performance you want, you might want to try wiping the cache and settings off your iPhone. This is how you do it with iOS 15:

- Go to *Settings*.
- Tap *General*.
- Tap *Reset*.
- Tap *Reset All Settings*.
- Enter your *passcode* if you have one enabled

Make sure you have your iPhone's Wi-Fi credentials handy before using this method to restore your phone to its factory settings. Re-entering them will be necessary.

Cleanup Your Storage

If you've owned your iPhone for a long time, there's a good chance you've accumulated a lot of unnecessary junk. You might be able to get a bigger iPhone if you delete this mess.

To begin with, you'll want to see how much space you have available on your device. This is exactly how you do it:

- Go to *Settings*.
- Tap *General*.
- Tap *Storage space & iCloud Usage*.
- Select *Manage Storage Space*

If you're getting close to the limit, head back to *Settings > General > iPhone Storage Space*. You'll be able to see all of your information here.

Apple can offer recommendations based on how you utilize your storage space, but you can also go through each area manually and delete data you no longer need.

Stop Auto Downloads

The automatic update feature of iOS is convenient, but it can slow down your iPhone's performance if it is constantly receiving updates.

Whether you don't mind manually upgrading your apps from the App Store, you might want to give disabling Auto Downloads a shot to see if it makes a difference in performance.

You'll need to go to Settings to accomplish this. Next, select iTunes & App Store from the menu. After that, you'll need to turn off Auto Downloads, which is located in the Auto Updates area. In addition, you may want to turn off other settings.

Disable the widgets.

Widgets got a significant facelift in iOS 15 thanks to a slew of design changes. However, if you don't use widgets on your iPhone, you may want to disable any or all of them to see if it makes a difference.

When you're on your iPhone's home screen, you'll need to slide to the right. A couple more choices await you after that:

- *Hard press* the *widget*
- At the bottom of the screen, click Edit. They may begin to wiggle as a result of this.

A little menu will appear if you press down hard on the widget. You'll need to touch on Delete Widget at the top of the screen to remove a widget.

If "*Edit*" was selected at the bottom, you can remove the widget by tapping the minus sign. You can even go all the way to the bottom and select Customize. Widgets can be readily added or removed using this method.

Widgets that you don't use should be disabled. Make sure to turn them back on if you don't notice an improvement in performance.

Clear Internet Browser Cookies & Data

Your phone's performance may be improved if you delete the cookies and other data stored by your browser.

In the browser settings app, touch Browser and scroll to where it says Clear Background and Website Data if you're using Apple's browser. It's just a matter of tapping it.

By pressing this button, you will be able to clear your browser's cache, cookies, and other data. iCloud-enabled devices can also have the background removed. Tap Clear Background and Data once again if you're okay with that.

Using Google Chrome, you'll need to touch the three horizontal circles on the right side of the app. If you haven't upgraded to Chrome's new features, they're in the top right corner.

To *clear browsing cookies, choose Settings, then Personal Privacy.* Now that you've made your selections, you can delete everything. Lagging can be alleviated by clearing the cache and restarting your computer.

Stop Using Background Rebrand-new

Using rebrand-new apps that function in the background by providing you with the most recent data when you open them is a bad idea in iOS 15. It also makes your phone work in the background, which you may prefer not to have.

Here is how to get rid of it:
- Open the *Settings application.*
- Tap *General.*
- Tap *Background App Rebrand-new.*
- Toggle *Background App Rebrand-new off* at the

top of the screen.

Reduce your list of apps and manually turn off the ones you don't use in order to keep them on for a few.

How to Repair Face ID Problems on an iPhone 12

In the event that your phone's Face ID feature isn't working, try the following.

To begin with, check to see if your iPhone 12 is running the most recent version of iOS on your device.

Having problems with Face ID on iOS 15? Check out the Face ID settings on the latest version of iOS.

- Navigate to the *Settings* menu.
- Take a look at *Face ID & Password*. Remember to use your passcode if you have one to get entry.

Make sure your mobile phone has *Face ID* set up and that the features you're trying to utilize *Face ID* with are switched on once you've entered the area you're in.

Keep an eye on the phone's display if you're having trouble unlocking it using your face.

In the event that you frequently alter your appearance, you may want to add a second face to your *Face ID* profile. Here's what you need to do to get a different look:

- Go to *Settings*.
- Tap *Face ID & Passcode*.
- Tap *Setup an Alternate Look*.

You should also check that your iPhone 12's front-facing camera is not blocked by any particles (dirt, etc.).

For best results, scan the face in a well-lit area if it isn't being registered by your device when you're setting up Face ID. Depending on where the iPhone 12 is located, you may also need to bring it closer to the person or remove it entirely.

How to Fix Overheating Issues on an iPhone 12

While using operating apps and service providers like *GPS*, we've seen reports of the *iPhone 12/Pro models* running hot during setup. The following are some options to consider if you don't need to take your phone into a store right away.

The first thing you should do is see whether deleting the offending app (*if you're using one*) makes a difference.

Try turning your cell phone off and back on as well. Even if it doesn't work, you can try placing the phone in *airplane mode*.

Free Bonus

Grab My *"Social Media Marketing Made Simple"* Ebook For **FREE!**

Today you can grab your copy of my Free e-book titled – **Social Media Marketing made Simple**. Best of all, it won't cost you a thing.

Click the image above to **Download the Book**, and also Subscribe for Free books, giveaways, and new releases by me.

https://mayobook.com/milton

Feedback

I'd like to express my gratitude to you for choosing to read this book, Thank you. I hope you got what you wanted from it. Your feedback as to whether I succeeded or not is greatly appreciated, as I went to great lengths to make it as helpful as possible.

<u>I would be grateful if you could write me a review on the product detail page about how this book has helped you. Your review means a lot to me, as I would love to hear about your successes.</u> Nothing makes me happier than knowing that my work has aided someone in achieving their goals and progressing in life; which would likewise motivate me to improve and serve you better, and also encourage other readers to get influenced positively by my work. <u>Your feedback means so much to me, and I will never take it for granted.</u>

However, if there is something you would love to tell me as to improve on my work, it is possible that you are not impressed enough, or you have a suggestion, errors, recommendation, or criticism for us to improve on; we are profoundly sorry for your experience (remember, we are human, we are not perfect, and we are constantly striving to improve).

Rather than leaving your displeasure feedback on the retail product page of this book, please send your feedback, suggestion, or complaint to us via E-mail to **"milton@mayobook.com"** so that action can be taken quickly to ensure necessary correction, improvement, and implementation for the better reading experience.

I want you to enjoy your reading experience; your satisfaction is my **#1** priority. You are well appreciated for reading this book.

Thank you, have a wonderful day!

About The Author

I'm a programmer, designer, and entrepreneur. I'm a full-stack developer and have been doing this for over seven years. I've worked for a few different start-ups and larger companies and am looking to start a new adventure. I love to learn, especially about technology and software. I like to build things and share my knowledge with others.

I don't care if you're a big company or a small business, I'm here to help you succeed.

I also enjoy the outdoors and hiking, so I can also offer some technical advice on that front. I've worked on a few different platforms and technologies including Node.js, HTML5, CSS3, and Sass, React, Redux, and Flux, Vue.js, Java, Groovy, and Grails, Ruby on Rails, WordPress, Magento, and Joomla, Django.

I want to make sure you're getting the most out of your devices, apps, and gadgets. I've worked with both large and small businesses. As a developer, I'm not just interested in development, but also in helping people use their devices to improve their life. I'm interested in solving problems and making sure people are getting the best out of their tech.

Subscribe to my Newsletter to download my Free Book, and also be informed about my new releases, and giveaways here: https://mayobook.com/milton

Connect with me on my Facebook Page here: https://fb.me/miltondonrandall

My Other Book(s)

I recommend these books to you, it will be of help, check it out after reading this book.

1. **3D Printer:** A Complete 3D Printing Guide
2. **iPhone 13 User Guide**
3. **How to Win Playing Cribbage/Draw Poker Card Games:** The Ultimate Guide on Rules & Strategies to Win and Beat the Odds Playing Card Games Like a Pro
4. **Draw Poker Handheld Game:** The Ultimate Beginner's Guide on Rules & Strategies to Win and Beat the Odds Playing Poker Card Games Like a Pro
5. **iPad Pro**
6. **iPhone 8:** The Complete User Guide is for Beginners
7. **iPhone 11:** The Complete User Guide for Beginners
8. **iPhone 12 User Guide:** The Complete New Guide to the iPhone 12 and iPhone 12 Pro Max, For Beginners and Seniors

INDEX

A
Android, 90
Apple, **108, 147, 171**
Apple ID, *147*
AR, 158, 159, 160

C
Calendar, 18
Customisation Options, 64
Customize, 66

E
Email, 15, 17

F
Face ID, 14

I
iCloud, 18, 20, **86**, *146*, **148**

iPhone 11, **11**, 12, 13, **104, 105, 148, 168**
iPhone 12, *104, 105*
iPhone 8, 149, 154, 158, 159, 160
iPhone X, **104, 105**
iTunes, **86**

R
Rest/Wake, **106**

T
Tips & Tricks, **168**

U
USB Limited Setting, **168**

V
Volume, **106**
Volume Up, *106*

CPSIA information can be obtained
at www.ICGtesting.com
Printed in the USA
BVHW020537290623
666543BV00010B/41